THE IMPROBABLE COMPUTER

How Quantum Computing Will Change the World

THE IMPROBABLE COMPUTER

How Quantum Computing Will Change the World

Giuliano Benenti
Università degli Studi dell'Insubria, Como, Italy

Giulio Casati
Università degli Studi dell'Insubria, Como, Italy

Simone Montangero
Università degli Studi di Padova, Italy

World Scientific

NEW JERSEY • LONDON • SINGAPORE • BEIJING • SHANGHAI • HONG KONG • TAIPEI • CHENNAI • TOKYO

Published by

World Scientific Publishing Co. Pte. Ltd.
5 Toh Tuck Link, Singapore 596224
USA office: 27 Warren Street, Suite 401-402, Hackensack, NJ 07601
UK office: 57 Shelton Street, Covent Garden, London WC2H 9HE

Library of Congress Control Number: 2026937680

British Library Cataloguing-in-Publication Data
A catalogue record for this book is available from the British Library.

THE IMPROBABLE COMPUTER
How Quantum Computing Will Change the World

ISBN 978-981-98-3240-8 (hardcover)
ISBN 978-981-98-3087-9 (paperback)
ISBN 978-981-98-3241-5 (ebook for institutions)
ISBN 978-981-98-3242-2 (ebook for individuals)

For any available supplementary material, please visit
https://www.worldscientific.com/worldscibooks/10.1142/14788#t=suppl

Desk Editor: Muhammad Ihsan Putra

Typeset by Stallion Press
Email: enquiries@stallionpress.com

Preface

> Every revolution begins with a simple question: what if the impossible were possible?

The race to produce increasingly powerful and fast computers has triggered an impressive miniaturization process. Thanks to this, today's computers can perform 100 million billion elementary operations (additions, multiplications, etc.) per second. But this miniaturization process cannot continue indefinitely: the size of transistors, getting smaller and smaller, is approaching atomic dimensions. Here, we enter the realm of quantum mechanics, where objects no longer obey the laws of classical physics, and computers as they are currently conceived would cease to function. We are faced with an insurmountable barrier, and the only way to bypass it is a complete change of strategy: building a new type of computer that follows quantum logic rather than classical logic—the quantum computer. This is not a mere improvement of current technology, but a paradigm shift, a profound revolution both in the logic by which computers operate and in the way information is transmitted from one point to another on Earth.

This fantastic possibility, illustrated in the first two chapters of this book, is based on a very simple difference: the fundamental building block of classical information, the bit, can exist in only two states—the state we call 0 or the state 1. The building block of the quantum computer, however, is the qubit, which can exist in any combination of the 0 and 1 states. If we imagine the classical bit as an object that can be only white or black, then the qubit is like an object that can take on all possible shades of gray.

This extraordinary flexibility of the qubit endows the quantum computer with possibilities absolutely unthinkable for a classical computer, no matter how powerful one might imagine building it. The future quantum computer could perform calculations in a few seconds that would take millions of years for the most powerful computer existing today. This is not a futuristic prospect but a not-too-distant reality.

The revolution sweeping the world of computers, i.e., information processing—is also forcefully entering the world of communication, namely information transmission. Today, every message, photo, or video travels through the ether from one point to another on the planet, relying on electromagnetic waves—from one cell phone to another, from one continent to another. These electromagnetic waves obey classical physics. But imagine a not-too-distant future where this infrastructure will be replaced by objects that follow quantum mechanics: photons. Photons are the fundamental unit of electromagnetic waves. We can think of them as massless particles that carry energy through space at the speed of light. This paradigm shift will offer enormous advantages. In particular, it will be impossible to intercept and copy a message without being detected. It will thus be possible to rebuild the information transmission system on new, inherently secure foundations. Not only will secret diplomatic communications be protected as never before, but even simple daily transactions, such as the use of credit cards, will enjoy impenetrable security. Progress in this field is rapid. The first satellite capable of conducting intercontinental encrypted teleconferences according to quantum protocols was launched into orbit by China in 2017. The extraordinary future of information transmission is described in the third chapter, where we explore how quantum mechanics is rewriting the rules of communication.

In the final two chapters of this book, we look beyond the boundaries of quantum computing and communication toward an extraordinary and promising future. This new phase of our technological adventure moves into a fascinating and unexplored territory, where quantum sensors open the door to possibilities that defy the imagination. In particular, in the fourth chapter, we delve into the vibrant world of quantum sensors—instruments so sensitive they can perceive the smallest energy fluctuations of an active volcano or the subtlest signals of a diseased cell. We will explore how these advanced devices are already beginning to transform fields such as medicine, where they could detect diseases before they become visible with current technologies. The applications are vast and the potential enormous, marking the dawn of an era where our ability to observe, interact with, and protect our world will be amplified in ways we can only imagine today.

In the 1940s and 1950s of the last century, when the first computers appeared, they were seen as enormous metal cogs, useful only for monumental tasks such as the development of hydrogen bombs. Few would have imagined that these machines would earn a place in daily life. Today, the reality is extraordinarily different: almost everyone has a computer and a cell phone—devices that keep us connected, informed, and entertained. From the depths of the digital ocean of the Internet to the swirling currents of social media, technology has become an almost natural extension of our existence. Already today, the first prototypes of quantum computers have emerged at the forefront of technology, marking the beginning of a new computational era. The road to their full realization is fraught with technical challenges, a true labyrinth of complexity and arduous obstacles. Nevertheless, enthusiasm is high: from prestigious universities around the world to multinational giants and agile startups, a fervor of research and investment is fueling this revolution. It is difficult to predict exactly where this turmoil of innovation will take us and what impact it will have on our daily lives. However, it is conceivable that the changes around us will happen before we realize it. The future is a blank page on which new quantum technologies are beginning to write powerful changes. Think of the areas where vast amounts of data are constantly generated that current computers are unable to process, and which could be usefully analyzed by a quantum computer. Or future quantum communication networks, which could evolve into a quantum Internet.

It is plausible that the development of quantum computers will become increasingly intertwined with the development of artificial intelligence in the future, paving the way for new great opportunities. Alongside these, ethical problems may arise related to the distinction between natural and artificial—the foundation on which the values of our society are built. We believe it is the responsibility of scientists to present the meaning of such discoveries in a way that is understandable to everyone. It is this conviction that motivated us to write the volume you hold in your hands.

Contents

Chapter 1

Inside the Invisible: A Journey into the Quantum Universe

> Anyone who is not shocked by quantum theory has not understood it.
>
> Niels Bohr

1.1 Discovering Indeterminacy: Probability in the Quantum World

The purpose of this book is to explain the workings of a quantum computer in a simple and understandable way, and then introduce readers to the world of quantum communication and other quantum technologies—the future that is upon us. The first step on this path is to familiarize ourselves with quantum mechanics, which has profoundly revolutionized our understanding of the world. We must take these initial steps cautiously, as quantum computers, like other future quantum technologies, are based on counterintuitive concepts that contradict common sense and our daily experience. For this reason, quantum mechanics has fascinated and continues to fascinate scientists and all of humanity for over a century.

The essential characteristic of this theory is that it forces us to abandon the certainties of classical physics and introduce an element of probability. It is not possible to predict with certainty the occurrence of a phenomenon; we can only calculate the probability of it happening. This new physics also takes on the character of a new philosophy: it influences and is influenced by art, music, literature, psychoanalysis, and the social sciences. During the period when quantum mechanics was born, Picasso painted *Les Demoiselles d'Avignon*, Freud wrote *The Interpretation of Dreams*, and Schönberg composed *Pelleas und Melisande.* It is important to remember how the first decades of the last century witnessed a shift in perspective from the past that affected various activities of the human intellect simultaneously, leading to the surpassing of the previous worldview and culture, later called "classical".

If a curious person were to ask a scientist expert in the field what quantum computers are, the first response they would probably receive is the simplest and most perfunctory one—avoiding having to delve into discussions about quantum parallelism, exponential growth, etc.: "Quantum computers are computers that use the rules of quantum mechanics to process information—perform calculations—and for this reason, they are enormously more efficient than normal classical computers". This response, which you will probably find in popular science articles on the topic, leaves the curious person at first glance with the feeling that they may have come closer to the answer they were looking for; upon reflection, however, they realize they have actually remained at the starting point. In fact, this definition simply shifts the problem. Why are quantum computers so powerful?

To answer this question, we must start with how we describe a quantum system; it is completely different from what we are accustomed to in our daily experience. In the world of classical mechanics, it is natural to observe the properties we are interested in for a given physical system and then assign them one or more values. For example, to describe a moving car, we focus on aspects such as speed and position, measure these quantities over time, and record them. This method not only allows us to know the car's current state but also to predict future events. For instance, if we are watching a Formula One Grand Prix, the fastest car will probably win.

Now imagine reducing the size of the racing car we are observing until it becomes as small as a few atoms. We will then begin to notice different behaviors. These few atoms will start to react to our measurement, making our lives extremely complicated! Why? How do we determine the car's speed? We must measure its position at two consecutive moments, and then, knowing the distance traveled and the time taken, we can obtain the speed. Thus, measuring speed reduces to measuring position, which is done with tools such as human eyes, cameras, or laser sensors-all devices that ultimately collect light or other forms of electromagnetic radiation emitted or reflected by the car. This measurement process seems natural and obvious to us, and scientists have taken it for granted for centuries, to the point where they never thought it could lead to conceptual difficulties.

In reality, this seemingly harmless process is at the core of the need to abandon classical physics and introduce new laws to describe atomic systems. In fact, if our car is the size of Formula One cars, we can measure its position without altering it or its speed: the light directed at it to detect its position will have practically no effect on the car. On the contrary, if the car is composed of a few atoms, the light needed for the measurement, even if extremely weak, will interact with the atoms themselves and have effects—for example, it will change the car's position by producing a push that alters its speed. Even a single photon, the smallest amount of light that can be emitted, will have an effect on an atom equivalent to firing a cannonball at a Formula One car to measure its position: this cannonball will certainly have an impact on the car! Therefore, each single impact of light can have a significant effect, changing the position or speed of the "atomic car" with meaningful consequences. This surprising reality is why we must reconsider and reformulate the laws of physics when we venture into the atomic world. We can no longer ignore that our observation tools interact with the observed object; in the quantum world, the very act of measuring alters what we are trying to understand.

This effect is the basis of the fascinating uncertainty principle, formulated in 1927 by Werner Heisenberg, a young 26-year-old German scientist who revolutionized our way of seeing the universe. The uncertainty principle, in fact, revealed an insurmountable limit: nature itself prevents us from knowing both the position and velocity of a particle simultaneously and with absolute precision. This discovery does not relate to the limitations of measurement tools, which could always be improved, but establishes a fundamental barrier inherent in the laws of the universe.

According to Heisenberg, any attempt to refine the precision with which we know one of these two variables necessarily increases the uncertainty of the other. No measurement tool, no matter how sophisticated we make it, will ever be able to overcome this limit. This principle not only places a limit on human knowledge but also opens a window into a world where uncertainty is intrinsic to the very nature of things. We also know the limit value for this uncertainty: the uncertainty in position multiplied by the uncertainty in velocity can never be less than the value of a universal constant, Planck's constant—a remarkably small number[1] that acts as an inviolable threshold. This mysterious number is not just a theoretical limit;

[1] In the International System of Units, Planck's constant takes a value of $6.62606957 \times 10^{-34}$ Joules · second.

it is an expression of the intrinsic structure of reality. As we strive to locate the position of an electron with increasing precision, its velocity becomes increasingly elusive, such that the product of the two uncertainties is always at least equal to Planck's constant. Heisenberg's principle appears not only in the measurement of position and velocity but also in many other pairs of measurable quantities, such as time and energy. The uncertainty in the measurement of energy multiplied by the uncertainty in the measurement of time cannot be less than a constant that is always the same: Planck's constant. These pairs of quantities, which cannot be measured simultaneously with arbitrary precision, are called "conjugate variables".

Planck's constant thus sets a limit to the precision of our measurements-a limit below which it is absolutely impossible to go. Since the numerical value of this constant is extremely small, this limitation is not perceptible in daily experience. The existence of this universal constant of nature and the reason why it assumes exactly that specific value remain one of the most fascinating mysteries of science.

The uncertainty principle forced the founding fathers of quantum mechanics to radically revise their conception of the universe. The indeterminacy of conjugate variables, in fact, obliges us to embrace a description of the universe in which at least one of the two conjugate variables does not have an exact value for fundamental reasons, not due to our ignorance. Forced to abandon the exact description of a system, physicists adopted a probabilistic approach.[2] In other words, we cannot predict with certainty what the result of a measurement will be, but only calculate the probability of obtaining one result rather than another.

In fact, in the classical world, we can assume, at least in principle, that we can determine all the characteristics of a system with increasing precision: position, velocity, size, weight, temperature, and any other property we wish to measure. Perhaps we cannot do it today, but in the future, by improving our tools, we will be able to obtain increasingly precise measurements. In the quantum realm, however, as we have seen, Heisenberg's uncertainty principle does not allow us to measure conjugate variables such as the position and velocity of a particle perfectly and simultaneously. Consequently, to describe the properties of a quantum system, we must adopt a probabilistic approach and thus introduce a probability distribution.

[2]For this purpose, they introduced a mathematical tool called a wave function, which does not tell us exactly where to find, for example, an electron, but allows us to calculate the probability of finding it in a given location.

Think of rolling an unbiased die: each face has an equal probability (1/6) of appearing. This simple example of a probability distribution in our macroscopic world gives an idea of how quantum mechanics describes the behavior of particles—not with certainties, but with probabilities. And just as we cannot predict which number will appear when we roll a die, we cannot predict the results of quantum measurements with certainty, but only calculate the probabilities of the various possible outcomes.

The probabilistic description is not only a distinctive feature of quantum mechanics but also a true paradigmatic leap compared to classical mechanics. Think of Newtonian physics: we can precisely measure the position and velocity of a cannonball at the moment of firing, and by applying Newton's laws, predict with absolute certainty the exact point where it will hit the ground. It is a world of crystal-clear certainties, where every cause has a predictable effect. But now we enter the subatomic domain, where the rules of quantum mechanics reign. Here, take an electron: we can no longer speak of definite positions or velocities. All we can do is measure the probability of where the electron might be and how fast it is moving at a given moment. Instead of predicting an exact result, we calculate a cloud of possibilities—a set of potential future positions, each with an associated probability. This cloud of possibilities evolves over time, like a wave of probability that expands and changes. When we finally measure the point where the electron impacts, for example, on a photographic plate, the specific result is the product of the collapse of this cloud into a precise point, but only at the moment we perform the observation. Quantum mechanics, therefore, does not tell us where the electron "will be", but where it "could be", transforming our understanding of reality.

The legend of William Tell, used by Max Born to explain the concept of probability in quantum mechanics, offers a fascinating metaphor. Imagine Tell, his heart pounding, ready to shoot an arrow at the apple placed on his son's head. Before acting, he whispers a prayer, asking God for a steady hand and an infallible aim. Similarly, a physicist in his laboratory directs a neutron at an atom, hoping to hit it precisely. He knows that with experience and improved tools, he can perfect the experiment more and more. However, unlike Tell, the scientist is aware that there is an insurmountable limit imposed by Heisenberg's principle. If the cruel bailiff Albrecht Gessler had forced Tell to aim not at an apple but at an atom on his son's head, using a neutron as an arrow, even Tell's extraordinary skill would have made no difference: the success of the shot would have been determined solely by the whims of probability.

It is not surprising that Heisenberg's uncertainty principle met resistance even from the greatest scientists of his time before being accepted. Among these, the most prominent opponent was Albert Einstein, who never accepted the idea of a universe governed by chance. His famous statement, "God does not play dice with the universe", reflected his rejection of the probabilistic aspect of quantum mechanics. God may be subtle, but he is not malicious, Einstein said, and for him, the idea that the behavior of the cosmos was guided by probabilities rather than deterministic and precise laws was unacceptable. Convinced that there was some hidden mechanism, a "hidden variable" capable of restoring order and predictability to the universe, Einstein devoted much of his life to trying to modify quantum mechanics to make it compatible with the determinism of classical physics. However, every attempt to uncover this mysterious alternative theory failed. In response to this stubborn search, Niels Bohr, one of the founding fathers of quantum mechanics, replied with an equally memorable phrase: "Einstein, stop telling God what to do". This exchange between two giants of science not only summarizes the conflict between determinism and probabilism but also marks one of the most fascinating and profound philosophical discussions in the history of physics.

With this first step, we have lifted the veil on one of the most fascinating aspects of quantum mechanics: the realm of probability. Now we are ready to discover how this revolutionary characteristic can transform the world of computing. What is a "quantum computer" really, and what incredible possibilities can it offer? To answer, we must start from the very foundations of information processing, from the fundamental building block on which every classical information processor is based: the bit. In the quantum realm, the bit has a much more powerful and versatile twin: the qubit. We will explore how this fundamental unit, guided by the laws of quantum physics, can open new frontiers in computing, leading us toward possibilities that until recently seemed pure science fiction.

1.2 From Bit to Qubit

The bit is the unit of information; it can take two values and can be represented by any system that can exist in two distinct states—for example, on or off, which we call 1 and 0, respectively. Imagine a coin that can land on heads or tails, a ball that can be in one box or another, or a sound pulse that can be long or short, as in Morse code. In today's computers, this bit is represented by an electric current that can flow through a wire or be blocked. At the center of this process is the transistor, a tiny device that acts as a switch, controlling whether the current flows or stops. In a classical computer, billions of these transistors work together, turning on and off in a coordinated dance that allows incredibly complex calculations to be performed. It is thanks to these tiny switches that computers can solve problems, process data, and power the digital world we live in.

The importance of the bit lies in its ability to represent any number, word, or message through a simple sequence of 0s and 1s, known as binary encoding. Think of numbers: in the binary world, each digit is represented as a sequence of 0s and 1s. For example, the number 2 becomes "10" in binary; 3 becomes "11", 4 becomes "100", and so on. But the beauty of binary encoding does not stop at numbers: every letter of the alphabet, every word, and even entire books can be translated into this language of zeros and ones. This simple yet powerful system is the foundation of the entire digital world, capable of converting any thought, data, or image into a binary code that computers can manipulate, transmit, and store.

The quantum bit, or qubit, is a much more sophisticated and powerful object than the simple classical bit. While the latter can exist in only one of two states, 0 or 1, the qubit can exist in both simultaneously, thanks to the fascinating phenomenon of superposition. To visualize this difference, imagine a globe: the two states of the classical bit are like the North Pole and the South Pole—fixed and opposite. In contrast, the possible states of a qubit are represented by all points on the surface of the globe, offering an infinite number of intermediate positions. But what does it mean to say that the qubit is in both states 0 and 1 at the same time? To begin to understand, think about the results of measuring the qubit. As mentioned earlier, the result of a measurement in quantum mechanics is probabilistic; we do not know if we will get 0 or 1. Before making a measurement, the qubit is generally neither at the North Pole (0) nor the South Pole (1), but in a combination (superposition) of the two states. If we look at the globe, each point on the sphere determines the probability with which we will find the qubit in 0 or 1 when we measure it. For points in the northern hemisphere, we will most likely find 0, while for points in the southern hemisphere, we will most likely find 1. A point at the equator corresponds to a state where the qubit has a 50% probability of being measured as 0 and 50% as 1, while a point closer to the North Pole might have a 70% probability of being measured as 0 and 30% as 1; a point exactly at the North Pole will be measured as 0 with certainty, and a point at the South Pole as 1. In the end, when a qubit is measured, it is "forced" to choose one of the two states, 0 or 1 (the North Pole or the South Pole). But before the measurement, the qubit is in a superposition of both, represented by a point on the globe. In summary: superposition means that the qubit can exist in a combination of both states (0 and 1) simultaneously until it is measured. The position on the sphere (the globe) represents this combination, which changes based on the point's position on the surface.

This means that the qubit operates in a much vaster universe of possibilities, where every point on the globe is a potential state of information. While the classical bit is limited to a simple transition between 0 and 1—like turning a switch on or off or moving from the North Pole to the South Pole in the globe analogy—the qubit opens the door to countless transformations, comparable to all the infinite paths that can be traced on a globe. Instead of being limited to moving from the North Pole to the South Pole (or vice versa), the qubit can take any journey on the sphere's surface.

This extraordinary richness of states and transformations, when working with many qubits together, promises to revolutionize the way we approach computing. Scenarios emerge where problems that today push the limits of the most powerful classical computers can be solved by quantum algorithms, harnessing the vast potential hidden in these tiny yet incredibly complex building blocks of information.

1.3 Questioning Uncertainties: Measurements in the Quantum Labyrinth

The qubit is a quantum creature, whose behavior is governed by the laws of quantum mechanics. And these laws allow us to determine all the probabilities of obtaining a certain result when we measure it. The possible states of the qubit can be represented by any point on the surface of a globe. This infinite possibility of states and consequently of transformations from one state to another is what makes the qubit such a fascinating and powerful component in the world of computing, capable of exploiting quantum rules that defy our classical intuition. However, when we finally observe it, the qubit "collapses" into one of the two classical states: 0 or 1, as if constrained to choose between the North Pole and the South Pole.

How can we truly grasp the concept of superposition in quantum mechanics, where an object can exist simultaneously in two distinct states, such as 0 and 1? And how does this differ from the more familiar concept of probability in classical mechanics? To fully understand, we must explore what happens at the crucial moment when we perform a measurement.

It is at that instant that the quantum world reveals its extraordinary nature and uncovers the fundamental difference between superposition and simple probability.

Imagine having a classical bit closed in a box. Someone before us has prepared it in a specific state: it could be 0 with a 50% probability, or 1, also with a 50% probability. Although we do not know the state of the bit, one thing is certain: the bit is definitely in one of these two states, 0 or 1. This is an element of physical reality, something concrete and indisputable, regardless of whether we decide to open the box or not. When we finally open it and observe the bit's state, we are not changing it; we are simply discovering a reality that already existed. The measurement process, in this case, does not alter the system—the bit—in any way, but provides information about it. It is a bit like observing a racing car speeding down the track: the observation does not affect its speed or position; it only provides information we did not have before.

In quantum mechanics, the situation changes drastically. Imagine having a qubit closed in a box. When we open it and take a measurement, we will find, just like in the case of the classical bit, that the qubit takes the value 0 or 1. However, unlike the classical bit, we cannot say that the qubit was already in one of these two states before the measurement. On the contrary, before opening the box, the qubit could have been in a state of superposition, represented by any point on the surface of the globe. This means that while in the classical world, if we discover that the bit is in state 0, we can definitively conclude that it was in the same state even before opening the box, in the case of the qubit, things are not so simple. If we measure the qubit and find 0, we cannot say that it was actually in state 0 before the measurement. In other words, we cannot consider the state of the qubit before observation as an element of physical reality. The measurement itself "decides" the result, and the state of the qubit before opening the box remains undefined and inherently uncertain.

An analogy to understand the quantum measurement process is offered by the famous young woman-old woman figure (Figure 1.1), which dates back to an 1888 German postcard. This image, which has captured the attention of psychologists for decades and is commonly known as Boring's wife/mother-in-law (after the American psychologist Edwin Boring), represents neither a young woman nor an old woman—at least not until we observe it. The image is inherently ambiguous: depending on who looks at it, at first glance, it may appear as an old woman with a large nose, a scarf, and a fur collar, or as a young woman with a necklace, a fur coat, and an elegant hat. Curiously, studies show that people under 30 tend to see the young woman first, while those over 30 usually identify the old woman first.

Fig. 1.1. Young woman or old woman? (William Ely Hill, American cartoonist and illustrator, 1915.)

In this analogy, the image represents a state of superposition, containing both possibilities: the young woman and the old woman. Only when the observer looks at the figure does the perception "collapse" into one of the two interpretations, determining which of the two women we see. This process of perceiving the image is similar to what happens in quantum mechanics, when measurement causes the "collapse" of a superposition state into a single defined reality.

1.4 From the Mona Lisa's Smile tu Qubits: The Art of Superposition

The image of the young woman and the old woman is only an analogy, useful for giving us a vague idea of what happens in the quantum world.

It helps us intuit how our minds, accustomed to classical logic, struggle to understand the concept of quantum superposition. An even more fascinating and complex example of superposition can be found in Leonardo da Vinci's Mona Lisa. The enigmatic smile of the Mona Lisa, which has enchanted and challenged humanity for centuries, is a perfect representation of this idea. Its charm stems from the fact that it eludes a precise definition: it is an intangible blend of happiness and sadness. One cannot think that Leonardo, with his profound knowledge of anatomy and optics, achieved this effect by chance.

Every time we observe the Mona Lisa, our perception changes imperceptibly, as if the painting were alive and changing before our eyes. But how did Leonardo create this sublime fusion of two such opposing emotions? We can imagine the Mona Lisa in a state of superposition between "happiness" and "sadness", like a qubit that can exist in both states 0 and 1 simultaneously. In fact, if we carefully analyze Figure 1.2, we find elements that evoke feelings of joy and melancholy in us. The shadow at the right corner of the mouth, sloping downward, suggests sadness. But further to the right, the shadow that curves upward, extending the line of the mouth in a way, resembles a smile. The shadows above the eyes make us perceive melancholy, while the light between the left eye and the temple gives a more lively expression. Figure 1.3 further clarifies the concept: on the left, the details that express "happiness" have been removed, leaving only "sadness" to emerge; on the right, the opposite. When we observe one of these two versions, we perceive a single emotion, thus reducing the complexity of the work and eliminating its magic. We can therefore think of the two panels in Figure 1.3 as representations of the two distinct states of the Mona Lisa system, while the original painting (Figure 1.2) is a superposition of happiness and sadness.

We are almost at the end of the journey that will allow us to introduce the extraordinary potential of quantum technologies and information. Only one last step remains. So far, we have explored superposition in the context of a single qubit, but a true quantum computer is made of many qubits. What happens, then, when not just one, but two or more qubits enter a quantum superposition? This is where we delve into one of the most fascinating and revolutionary concepts of quantum physics: entangled states—a world where connections between particles go beyond all imagination and create possibilities that further challenge our understanding of reality.

Fig. 1.2. The smile of the Mona Lisa.

1.5 Invisible Threads: Quantum Entanglement

The possible states of two classical bits are four: 00, 01, 10, and 11, and at any given moment, they can exist in only one of these four states. There are no other possibilities.

With qubits, however, we enter a completely different world. Since they are governed by the laws of quantum mechanics, two qubits are not confined to just one of these four states. Analogously to the case of a single qubit, they can exist in a superposition of all these states, as if embracing every possibility simultaneously. This unique ability opens the door to a surprising phenomenon with no equivalent in the classical world: entanglement. It is a profound connection, an "intertwining" between qubits, that allows them to behave in a correlated way even at great distances, as if magically synchronized.

Erwin Schrödinger, one of the great pioneers of quantum mechanics, defined entanglement as "the characteristic trait of quantum mechanics". This statement highlights how fundamental it is to fully understand what entanglement is and why it is so different from the concept of superposition, which we can also find in the world of our daily experience. Superposition, in fact, is not exclusive to quantum physics. Think, for example, of the string of a musical instrument. When struck on a piano, plucked on a guitar, or bowed on a violin, the string produces sounds that our ear distinguishes

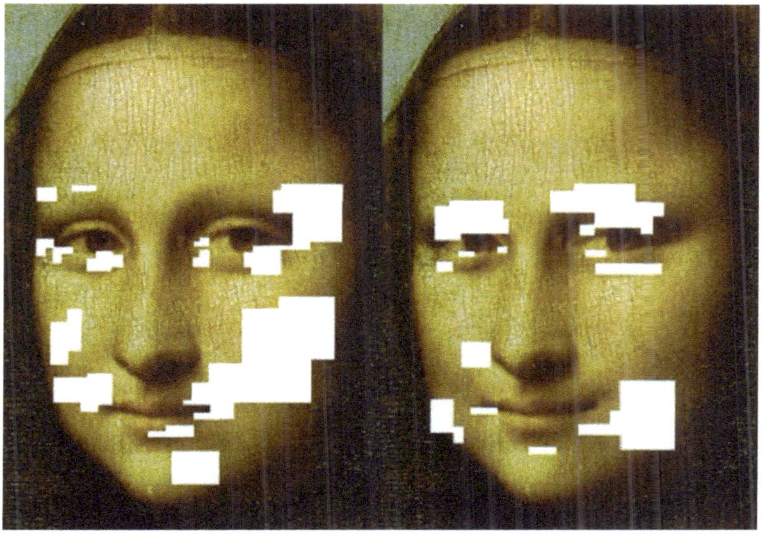

Fig. 1.3. Sad Mona Lisa (left) and happy Mona Lisa (right). The original painting of Figure 1.2 owes its magic to the "superposition" of these two "states" of the Mona Lisa. Figure is taken from S. Prvanović, "Mona Lisa - The Ineffable Smile of Quantum Mechanics", `https://arxiv.org/pdf/physics/0302089`.

between low and high based on their frequency. Low sounds correspond to low frequencies, high sounds to high frequencies. However, the sound produced is never a single pure frequency: it is a superposition of many frequencies that blend together. It is precisely this combination of frequencies that gives each instrument its unique timbre, allowing us to recognize it among thousands of others.

Now imagine taking two different musical instruments, each with its own distinctive sound. These sounds do not influence each other: they remain separate and independent. In the classical world, reality works this way. But in quantum mechanics, something extraordinary happens: we can create superpositions between the states of two different objects, as if they were mysteriously intertwined. Let us try to make this more concrete. Suppose we consider two people and focus on one of their characteristics: eye color. In our common experience, eye color is an objective fact independent of the observer. Whether someone looks at them or not, a person has light eyes or dark eyes, and this is a fact of reality.

But quantum mechanics turns this concept upside down. According to this theory, the two people can be prepared in an entangled state, where their eyes are in a superposition: a combination of the state "light eyes for the first person, dark eyes for the second" and "dark eyes for the first person, light eyes for the second". This is where quantum entanglement comes into play. Now imagine that one of the two people goes to an observer we will call Alice, while the other visits an observer named Bob. Alice could be in Europe and Bob in the United States, separated by thousands of kilometers, yet quantum mechanics establishes that their observations are deeply connected. When Alice looks at the eyes of the person in front of her, she may see light or dark eyes with equal probability; we cannot predict their color with certainty. But at the moment Alice observes them, something incredible happens: the eyes of the person observed by Bob, far away, become instantly defined. If Alice sees light eyes, Bob will definitely see dark eyes, and vice versa, as if synchronized by an invisible force. This is because the states of the two people are entangled. And just like with two entangled particles, acting on one immediately influences the state of the other, regardless of the distance. It is as if, at a deep level, these two people are a single entity, connected by an invisible thread that defies our understanding of the world.

Naturally, we cannot really quantum-entangle two people as in the previous example. However, what is impossible for us on a macroscopic scale becomes absolutely real in the microscopic world. It is indeed possible to create and observe entangled states between particles such as photons or atoms. These particles, although infinitely small, exhibit quantum correlations that go beyond what is possible in classical systems: their invisible and mysterious connections take us into a territory that challenges our intuition, derived solely from experience of the classical world. All this will become clearer when we explore Bell's inequalities, the key to unraveling the enigma of entanglement.

1.6 The Nature of Correlations

Schrödinger himself was aware of the strangeness of entanglement and its effects even at great distances. In 1935, he wrote a letter to Albert Einstein, in which he commented on entanglement saying: "I do not like it, and I am sorry I had anything to do with it". Schrödinger, despite being one of the founders of quantum theory, was fascinated but at the same time troubled by the conceptual implications of entanglement. Einstein called it "spooky action at a distance", expressing his discomfort with the idea that one particle could instantly influence another, regardless of the distance separating them.

1.6.1 *Classical correlations*

Two systems are said to be correlated when knowledge of a property of one provides information about the other. The concept of correlation also exists in the classical world, and Einstein himself used the metaphor of a pair of gloves to explain it. If we put the two gloves in separate boxes and send one to a distant person, opening this box will automatically reveal which glove is in the other, regardless of their distance. This type of correlation is completely deterministic—there is nothing mysterious about it: discovering the right glove does not influence the left glove in any way, but only provides information about it.

Suppose now we have white balls and black balls available. We randomly choose a pair of balls of the same color and close each ball in a different box; then we send one box to Alice and the other to Bob. Neither of them knows if the ball in their box is white or black. But the moment Alice opens her box and discovers, for example, a white ball, she immediately knows that Bob also has a white ball, because the balls are the same color. In this case, Alice's action—opening the box and seeing the ball's color—allows her to predict with certainty the result Bob will get when he opens his box. This happens because the two systems (the two balls) are perfectly correlated: knowledge of one provides information about the other. Now imagine modifying the experiment just described: for each box, we randomly choose the color of the ball we put in it. In this way, the boxes sent to Alice and Bob can contain balls of the same color or different colors with equal probability. Therefore, after opening her box, Alice can deduce nothing about the color of the ball in Bob's box: the two systems are completely uncorrelated.

This simple example clearly illustrates how crucial correlations are: having correlated systems (such as two balls of the same color) allows us to obtain information about one part of the system by observing the other. But the real question is: how can we measure and quantify these correlations? The answer is not at all simple. In reality, systems are rarely perfectly correlated or completely uncorrelated as in our ball example. Correlations in the real world are more complex, nuanced, and often hidden behind a veil of uncertainties. This is why quantifying them opens the door to a deeper understanding of the very nature of connections between things.

Imagine repeating our experiment many times: each time we send a box to Alice and one to Bob, who open them and record whether they received a white or black ball. As shown in Figure 1.4, there are at least two possible opposite scenarios corresponding to the upper and lower sequences.

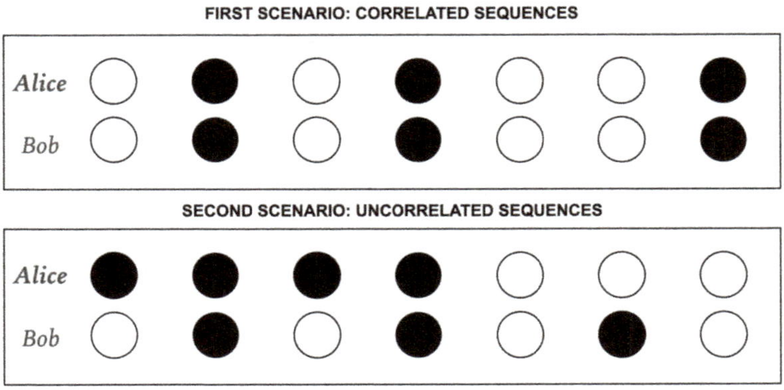

Fig. 1.4. Examples of perfectly correlated random sequences (top) and completely uncorrelated random sequences (bottom).

First scenario: each time we choose a pair of balls of the same color (white or black) and send one to Alice and one to Bob. Both see a random sequence of colors, but when they compare their sequences, they find that they are perfectly identical—their observations are correlated.

Second scenario: we choose a random sequence for the colors of the balls sent to Alice and a different, always random, sequence for the colors of the balls sent to Bob. In this case too, Alice and Bob see a random series of colors. However, when they compare their sequences, they find that some colors match while others do not, revealing that their sequences are completely uncorrelated. These two scenarios represent the extreme cases: from perfect correlation to a complete absence of correlation. Now, if on some occasions we sent balls of the same color to Alice and Bob, and on other occasions balls of randomly chosen colors, their sequences would be only partially correlated.

Based on these observations, mathematical algorithms have been invented to quantify the degree of correlation—to assign a number that measures the degree of correlation between Alice's and Bob's color sequences.[3]

[3]To quantify the degree of correlation between Alice's and Bob's color sequences, we can proceed as follows. Start from zero and add one unit each time Alice and Bob see the same color, and subtract one unit when the two colors are different. In the case of

So far, we have given an idea of what correlation means in the classical world, but we have not yet touched on quantum mechanics. The question now is: what is the difference between classical and quantum correlations?

1.7 Entanglement and the Reality of the Physical World

Quantum correlations lead to consequences that deeply challenge our intuition and are completely unacceptable for the way we are accustomed to thinking about the world. An extraordinary example is the famous EPR paradox, devised in 1935 by Albert Einstein, Boris Podolsky, and Nathan Rosen. Einstein himself proposed this "thought experiment" with the intention of showing that quantum mechanics leads to absurd conclusions and therefore must be modified. The EPR paradox, in fact, calls into question one of the fundamental pillars of our thinking: the principle of reality. This principle, in simple terms, states that reality exists independently of whether we observe it.

For example, if we open the door to a room and see a table, the principle of reality assures us that the table was already there, even before we opened the door and observed it. More formally, if measuring a physical quantity of an object (for example, its position or velocity) gives us a certain result with certainty, then that result is an element of physical reality, true regardless of our measurement.

This principle is considered obvious and indispensable. However, what Einstein, Podolsky, and Rosen demonstrated—and what makes their thought experiment so paradoxical—is that the principle of reality is in direct conflict with quantum mechanics. The two cannot coexist. We are therefore faced with a radical choice: either we abandon quantum mechanics, or we question the principle of reality itself—a concept that seemed unshakable in our understanding of the world. This is what makes the EPR paradox so fascinating and destabilizing.

We cannot elaborate on its content here, but through an example we can give an idea of this paradox, which led Einstein to consider quantum mechanics an inappropriate theory for describing nature. Take a stamp and

the first scenario, perfect correlation—when the colors of Alice's and Bob's balls are the same—we get a number equal to the length of the sequence. In the second scenario, where the colors are completely uncorrelated, we will sometimes add one unit and sometimes subtract it, remaining close to the starting point (zero). An intermediate value between those obtained in the two ideal cases of perfect correlation and complete uncorrelation indicates and quantifies a partial correlation.

cut it into two parts, either vertically or horizontally (Figure 1.5). Send one half to Alice and the other to Bob, who are very far apart, so that their observations cannot influence each other. Suppose Alice observes her half first. If she discovers she has the left part of the stamp, then we know with certainty that Bob will have the right part. If, instead, Alice has the right part, then Bob will have the left one. This is an example of classical correlation: the result of one measurement (the half of the stamp) is perfectly correlated with that of the other. The same applies if we cut the stamp horizontally: if Alice sees the upper part, Bob will have the lower part, and vice versa. It is important to note that the two cases (right/left or top/bottom) are mutually exclusive.

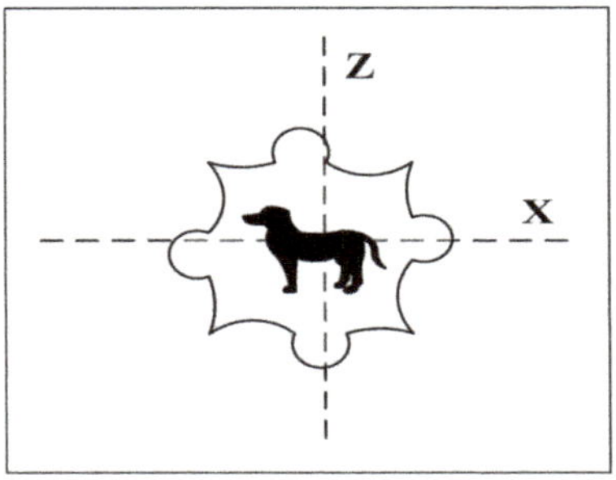

Fig. 1.5. Horizontal/vertical cut of a stamp (courtesy of Sara Felloni).

But in quantum mechanics, the situation is very different. It is as if Alice, after receiving her half of the stamp, could choose, at the moment she observes it, whether the stamp was cut horizontally or vertically. This means that her choice to measure the stamp determines what Bob will see, regardless of the distance separating them. If Alice decides to observe in terms of right/left, then Bob will see the opposite part; if she measures top/bottom, Bob will see the complementary part. This situation seems completely nonsensical from a classical point of view, where we expect that before any observation, Alice and Bob already possess their halves—right/left or top/bottom. In quantum mechanics, however, thanks to the phenomenon of entanglement, the result of Alice's measurement instantly influences Bob's result. It is this "instantaneous" interaction, far from our classical intuitions, that leads to paradoxical conclusions that are so difficult to accept.

1.8 Bell's Inequalities: At the Boundaries of Reality

Einstein was firmly convinced of a universe where reality exists independently of our observation, and he never accepted the deeply counterintuitive characteristics of quantum mechanics, such as those highlighted by the EPR paradox. For Einstein, physics should describe an objective world, not a game of probability governed by the observer. This vision is brilliantly captured in a famous letter he wrote to the physicist Niels Bohr, who defended the probabilistic interpretation of quantum mechanics—now widely accepted. Einstein wrote:

> Quantum mechanics is certainly imposing. But an inner voice tells me that it is not yet the real thing. The theory says a lot, but does not really bring us any closer to the secret of the Old One.

In this phrase, the "Old One" represents the universe itself, enigmatic and hidden behind the veil of the probabilistic appearances of quantum theory. Einstein perceived quantum mechanics as incomplete-a path that offered practical answers but did not reveal the ultimate truths of the universe. This profound divergence of vision gave rise to a controversy that lasted for decades: on one side, Bohr, Heisenberg, and other supporters of the orthodox interpretation; on the other, Einstein and Schrödinger, who refused to give up a deterministic vision of the world. This philosophical debate, in addition to being a confrontation between giants of science, marked one of the most fascinating intellectual battles in the history of physics, pushing the brightest minds to explore the boundaries of reality and question the profound meaning of nature.

A decisive step was taken in 1964 by the Irish physicist John Bell, who found a way to bring the controversy from the philosophical sphere directly into scientific laboratories. Bell proposed an experiment that could resolve the issue of the EPR paradox and the nature of reality once and for all. Starting from the assumption that the principle of reality—the idea that reality exists independently of the observer—was valid, Bell developed mathematical relations today known as Bell's inequalities.

The extraordinary importance of these inequalities lies precisely in the fact that they could be experimentally verified. In other words, it was possible to conduct experiments with atomic systems, which follow the laws of quantum mechanics, to see if they obeyed Bell's inequalities or not. If the results had confirmed the inequalities, this would have meant that even atomic systems obey the principle of reality, and consequently, quantum mechanics would have had to be revised, considered incomplete or partial. But if atomic systems had violated Bell's inequalities, the conclusion would have been shocking: we would have had to abandon the principle of reality, accepting that the microscopic world does not behave according to our expectations and that entangled particles break the rules of our classical intuition. A result both fascinating and disturbing, which would have shaped the fate of our understanding of nature itself.

A simple experiment shows how a Bell inequality works. Imagine having a group of people and dividing them based on three characteristics: for example, male or female, under forty or over forty, and light or dark eyes. If we do a simple count, we see that the number of people—male or female—under forty with light eyes (dashed region in the upper part of Figure 1.6) is less than or equal to the sum of the number of males under forty (regardless of eye color) and females with light eyes (regardless of age) (dashed region in the lower part of Figure 1.6). This sum, in fact, includes people under forty with light eyes, plus males under forty with dark eyes and females over forty with light eyes.

We can verify that this inequality is always true for any group of people we consider. This happens because people's characteristics, such as gender, age, and eye color, are elements of reality that exist independently of whether we observe them.

On the other hand, in quantum mechanics, we can prepare entangled states that, as predicted by the theory itself, violate Bell's inequalities. In 1981, Alain Aspect, Philippe Grangier, and Gérard Roger conducted pioneering experiments that clearly demonstrated this violation. Quantum predictions, which challenge the so-called "principle of reality", proved correct! Using entangled states of photons—the fundamental particles of light and electromagnetic radiation—the results of these experiments marked an epochal turning point. It is interesting to note that Aspect was awarded the Nobel Prize in Physics only in 2022, more than forty years after his discovery. It may seem strange given the significance of the result, but it is certainly not the only case of belated recognition; for example, the Indian physicist Subrahmanyan Chandrasekhar had to wait more than fifty years to receive the Nobel Prize for his studies on stellar evolution, which paved the way for the definition of the black hole concept. Sometimes genius is so far ahead of its time, and the results obtained are so outside the ordinary, that they require a long time to be recognized.

Over the years, other laboratories around the world have replicated and confirmed Aspect's and his collaborators' results, consistently showing that

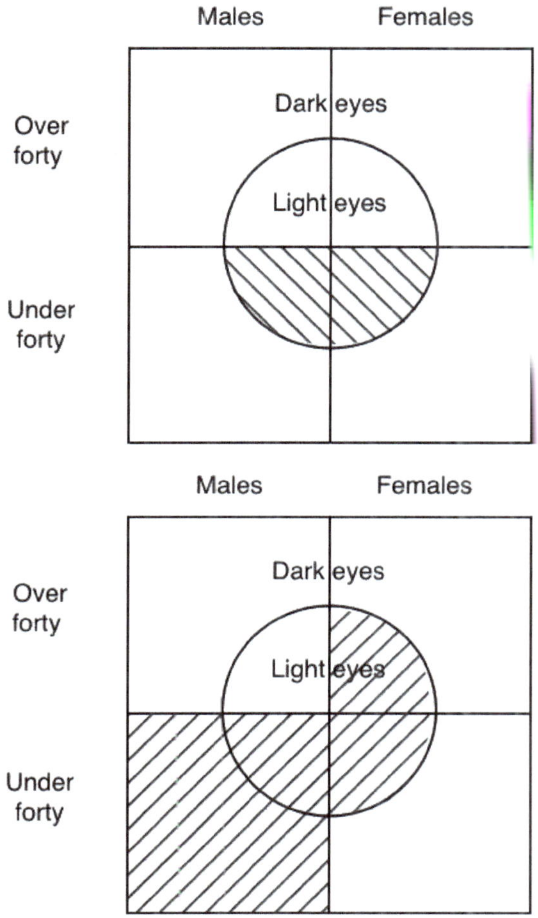

Fig. 1.6. Example of Bell's inequality in everyday life.

Bell's inequalities are violated, in perfect agreement with the predictions of quantum mechanics. This body of experimental evidence has led to a surprising conclusion: Einstein's deterministic vision of the world has been disproven. The experiments have resolved the dispute between the principle of reality and quantum mechanics, ruling in favor of the latter. Consequently, we must abandon the principle of reality and the concept of a rigidly deterministic universe, accepting the deeper and more mysterious nature of quantum reality.

In other words, we are accustomed to imagining that there exists a universe independent of us—a world we can study without altering it. Of course, we know that every observation or measurement interferes in some way with what we are observing, but we tend to believe that this disturbance can be reduced to negligible levels. Quantum mechanics, however, teaches us that this vision is an illusion. We can no longer think of the existence of an objective world, separate from our knowledge process and immune to our interactions. On the contrary, reality itself, at the quantum level, is intertwined with the act of observation. It is not a machine that proceeds mechanically, indifferent to our presence; what we obtain from nature depends on the questions we choose to ask and how we decide to observe the world. Quantum mechanics reveals a universe much more fluid and interactive than we could ever have imagined.

An illuminating example is the game of 20 Questions. Imagine an evening with friends, who after dinner decide to play the following game: one of them leaves the room, and the others agree on a secret word. At this point, the person who left re-enters and begins to ask questions: "Is it a living being?"—"No". "Is it on Earth?"—"Yes", and so on, trying to narrow down the field until they discover the mysterious word. The challenge is to guess it within 20 questions; otherwise, the game is lost.

It is our turn to leave the room. We stay outside for what seems like an eternity, until they call us back in. Upon re-entering, we immediately notice our friends' smiling and mischievous faces, as if they are plotting a joke. Despite this, we start the game as usual with our questions. At first, the answers come quickly: "Yes" or "No". But then something strange happens: the answers start to take longer and longer, and it seems bizarre, given that it's just a simple "Yes" or "No". After a few more questions, frustrated by the growing slowness, we finally make a guess: "Is the word 'cloud'?"—"Yes", the person we asked replies, and suddenly everyone bursts into loud laughter. Confused, we ask for an explanation. Here's the revelation: while we were outside, they had decided not to choose a word at all! Each person, before answering "Yes" or "No", had to mentally think of a word that was consistent with the answers given by the others up to that point. This is why the answers became increasingly slow: each new question complicated their mental game. It was no longer a matter of discovering a pre-chosen word, but of "creating" a hidden word.

These two versions of the game can be paralleled with the vision of classical physics and that of quantum physics. In the first case, reality is seen as something objective that exists independently of our observation. Just like the word already agreed upon by friends, the atom exists with a well-defined position and velocity, which we can simply discover. In the second case, however, the unknown word did not pre-exist: it emerged gradually, question after question, just as information about the atom takes shape only through the measurements the observer decides to perform. Furthermore, if we had asked different questions, the final word would have been different—exactly as the experimenter, by changing the type of measurement, would obtain different results. In this quantum vision, reality is not a simple photograph of the world, but a dynamic creation that depends closely on how we choose to observe it.

The comparison between the game and quantum mechanics, although not perfect, illuminates a crucial point. Quantum mechanics forces us to set aside the principle of objective reality, revealing that it is the very act of observing that defines what we call "reality". Similarly, in the game, the word "cloud" was not decided until it was gradually built by our questions. In both cases, it is the interaction between the observer and the phenomenon that shapes the world we perceive: what appears exists because we chose to investigate it.

But there is an extraordinary aspect we want to talk about now: what seems paradoxical in quantum mechanics—entanglement and quantum correlations—is not just a mystery, but an extremely powerful resource. We will see how these phenomena open new possibilities for computing, processing, and transmitting information—possibilities completely unreachable for classical physics. The fact that experiments demonstrate the violation of Bell's inequalities means that there are quantum correlations much "stronger" than those the classical world can offer. And since correlations quantify the information we can obtain about one system by measuring another, the stronger they are, the more information we can extract. This enhanced capability can be exploited to improve and expand our information transmission and computing technologies, going far beyond the limitations of classical methods.

The challenge now is the development and experimental implementation of algorithms that can fully exploit these potentialities. This is a field still in evolution, a formidable open frontier. We are now immersed in the depths of quantum mechanics, and with this, we are approaching the extraordinary technological horizons it offers—from computing to communication. Only now can we truly appreciate the revolutionary impact of these discoveries, capable of radically changing the way we conceive the reality around us.

Before delving into these topics, it is worth pausing on a surprising connection between games and quantum mechanics. This is not about using a game to explain quantum concepts, as we did in the previous paragraphs, but about imagining how games themselves could transform if we agreed to play following the rules of quantum mechanics instead of the classical ones we have been accustomed to since childhood. The dynamics and strategies would change, revealing a world of possibilities that challenge our expectations.

1.9 Quantum Games

Game theory is a fascinating discipline that explores mathematical models to analyze the dynamics of interaction between two or more decision-makers, called "players". Through a rigorous mathematical approach, it seeks to explain how and why certain behaviors and strategic choices can lead to the achievement of specific objectives. In other words, game theory reveals what the rational strategy is to adopt once the rules of the game are known. This ability to model decisions and interactions makes game theory crucial in a wide range of sectors, including economics, politics, and psychology, helping us better understand and predict the dynamics that influence human and social behavior.

Readers may wonder what the connection is between quantum mechanics and game theory. The latter, in fact, analyzes interactions between two or more players, each seeking to maximize a payoff or minimize a loss. At this point, it is natural to ask: what would happen if we applied the principles of quantum superposition and entanglement to players' strategic choices? In other words, what would happen if we extended game theory to the realm of quantum mechanics? Would a player who exploits the laws of quantum mechanics have an advantage over one confined to classical moves? And if all participants used quantum strategies, could they obtain better payoffs? Could new situations emerge where a "quantum advantage" arises? These questions open perspectives on how the integration of quantum physics and game theory could revolutionize our understanding of strategic and cooperative interactions.

The advantage offered by the principle of quantum superposition was brilliantly illustrated by an example proposed in 1999 by David Meyer of the University of San Diego. Imagine the famous spaceship Enterprise from the Star Trek TV series facing an imminent catastrophe. Q, the mysterious alien with almost unlimited powers and antagonist of Captain Picard, suddenly appears on the bridge and offers to help—on the condition that Picard can beat him in a challenge. The challenge is simple: Picard must place a coin with the "heads" face up. Then Q, Picard, and Q again, in turn, can decide to flip the coin or leave it as it is, without ever seeing it or knowing the opponent's moves. Q wins if, at the end of the game, the coin is still heads up; otherwise, Picard wins.

At first glance, it seems obvious that with an unbiased coin, Captain Picard would have a 50% chance of winning and saving the ship. Even with an extra move, Q gains no advantage if he does not know whether Picard flipped the coin or not. However, what would happen if Q, unlike Picard, could exploit quantum mechanics? The situation would change drastically. In this scenario, the coin becomes a qubit—a quantum object—and Q, as his first move, prepares a superposition of the heads and tails states, present in equal measure. Picard, in turn, can choose to do nothing or flip the coin. However, even if he decides to flip it, the effect would be nul : heads would become tails and tails would become heads, but the superposition would remain unchanged—both states, heads and tails, remain present in equal measure. At this point, Q makes his final move, reversing the operation that created the superposition and thus returning the coin to the heads state with absolute certainty, regardless of Picard's choice. In this way, Q ensures his victory, demonstrating the quantum advantage in a situation where, in the classical world, the result would have been uncertain.

In conclusion, Q always wins, without cheating—in the sense that he does not need to spy on Picard's move. However, the game is clearly unfair because only Q has access to the power of quantum mechanics. While Picard is confined to classical moves, Q can exploit strategies that are inaccessible to Picard, creating a quantum superposition of states. To illustrate, imagine that Picard, starting from the North Pole, can only choose to stay there or move to the South Pole, while Q is free to move across the entire Earth's surface: his first move takes him to the equator, and the last brings him exactly where he needs to be to win—back to the North Pole. This example clearly illustrates the superiority of a player who can use quantum strategies over one confined to classical logic, highlighting the fundamental advantage offered by quantum mechanics.

A classic and more complex example of the application of game theory is the famous "prisoner's dilemma". Imagine that the police have arrested two accomplices, Alice and Bob, suspected of committing a robbery. The evidence is sufficient only to convict them of one year in prison for illegal possession of weapons, while it is not strong enough to pin the 10-year sentence for the robbery on them. The two prisoners are held in separate cells, unable to communicate, and know exactly what the rules of the game are. If Alice decides to cooperate with the police and Bob chooses to remain silent, Alice will be released without any penalty, while Bob will be sentenced to ten years. If, instead, both decide to cooperate, they will be sentenced to an intermediate penalty of three years each. Finally, if both choose to remain silent, the police having no sufficient evidence to incriminate them for the robbery, they will receive the lightest sentence: one year in prison for illegal possession of weapons. This dilemma highlights the tension between individual interest and the common good, as well as how trust (or lack thereof) can influence strategic behavior.

In this game, optimizing the payoff means minimizing the number of years in prison. If the prisoners could cooperate, the ideal choice would be for both to remain silent, thus obtaining a light sentence of one year. This decision represents what economists call a "Pareto equilibrium", a concept introduced by the Italian Vilfredo Pareto in his *Cours d'économie politique* (1896). A Pareto equilibrium describes a situation where no one can improve their condition without worsening that of others, as in a system of perfectly competitive markets. In other words, it is the equilibrium point where all participants obtain the maximum possible benefit without compromising the interests of others. However, in the prisoner's dilemma, the temptation to betray the other makes this equilibrium difficult to achieve, creating tension between personal and collective interest.

Now imagine that Alice and Bob do not trust each other and seek to maximize their own advantage in a non-cooperative way. In this context, the rational move leads to the "Nash equilibrium", a strategy where no player can improve their payoff by unilaterally changing their decision. To obtain a better result, it would be necessary to act together, cooperating. But if we put ourselves in Alice's shoes, we can see the dilemma clearly. If Bob decides to confess, the best choice for Alice is to do the same: if she does not, she will be sentenced to ten years in prison instead of three. If, instead, Bob chooses to remain silent, Alice has every incentive to confess, as she will be released immediately. The same applies to Bob. The game is perfectly symmetric. The Nash equilibrium, therefore, inevitably leads both to confess, with the consequent sentence of three years in prison.

However, this result is less advantageous than the Pareto equilibrium, where both would remain silent and be sentenced to only one year of imprisonment. The problem is that the Pareto equilibrium requires cooperation—a behavior difficult to maintain in a game dominated by distrust.

Alice and Bob are thus faced with a real dilemma: rational logic pushes them to confess, even though silent cooperation would be more beneficial for both. This conflict between individual and collective interest is the heart of the prisoner's dilemma, where the fear of being betrayed guides every decision.

John Nash, the brilliant American mathematician who became a legend thanks to the movie *A Beautiful Mind*, revolutionized game theory with his insights. His studies forever changed our understanding of strategies and decision-making equilibrium, and for this extraordinary contribution, he was awarded the Nobel Prize in Economics in 1994. Nash demonstrated how individual rational behavior can lead to unexpected solutions, opening new paths in sectors far beyond economics, influencing politics international relations, and even biology.

The quantum version of the prisoner's dilemma, proposed in 1999 by Jens Eisert, Martin Wilkens, and Maciej Lewenstein, introduces a completely new dynamic compared to the classical game. In this scenario, the two prisoners share a pair of qubits—one qubit is in Alice's possession and the other in Bob's. As in the classical game, the two prisoners cannot communicate in any way. Their possible moves are now transformations ("journeys on the globe") for their qubit. In the end, Alice and Bob measure their qubits and obtain, with a certain probability, the result "confess" (state 0) or "not confess" (state 1). The objective remains the same: to optimize the payoff, i.e., minimize the prison sentence. However, thanks to the power of entanglement and quantum transformations, the players now have new strategies to influence the result, overcoming the limitations imposed by classical logic. This quantum version of the prisoner's dilemma opens completely unprecedented scenarios, where cooperation and betrayal intertwine with the laws of quantum mechanics.

We cannot go into the details of the calculations needed to optimize the game, but we can focus on the most important conclusion. If the initial state of the qubits has a sufficient level of entanglement (which, it is worth noting, can be quantified), and only one of the players—say Alice—is able to use quantum moves, then Alice will have a clear advantage, obtaining a better payoff, just like Q in the coin game. However, if both players can exploit quantum strategies, something extraordinary happens: above a certain level of entanglement, the Nash equilibrium also becomes an optimal Pareto equilibrium. This means that the best possible payoff in the classical cooperative game becomes achievable even in the quantum game, without the two prisoners needing to communicate or agree on a strategy. In other words, allowing players to use quantum strategies not only improves their possibilities but also completely eliminates the prisoner's dilemma. Quantum mechanics not only transforms the game but resolves the conflict between personal and collective interest, making successful cooperation possible without the need for trust or communication.

In addition to the two examples discussed above, many other quantum games—with two or more players—have been proposed and continue to be developed. These games are not only a fascinating pedagogical tool to refine our understanding of the counterintuitive mechanisms underlying quantum mechanics but also represent a potential source of inspiration for the development of software for future quantum computers. In this way, quantum game theory not only expands our theoretical horizons but could also play a crucial role in shaping tomorrow's technology.

1.10 Interlude: Order and Complexity

So far, we have discussed how probability is an intrinsic element of quantum physics, in stark contrast to the determinism of classical physics. However, the following pages aim to warn readers that determinism does not necessarily equate to predictability. Even in classical physics, uncertainty and probability play a significant role and often dominate the overall picture. In fact, knowledge of deterministic laws does not always guarantee the ability to predict the future with certainty: there are situations where the behavior of a system is unpredictable, despite the presence of deterministic laws.

The primitive world was perceived as total chaos, where every aspect of life—from human events to natural phenomena—was attributed to the will of capricious and unpredictable spirits. People saw the universe as a place dominated by uncertainty, where mysterious forces acted without an apparent logic. As described by John Milton in Paradise Lost (Book II), the universe was a realm where "eldest Night and Chaos, ancestors of Nature, hold eternal anarchy amid the noise of endless wars, and by confusion stand".

Subsequently, the observation of the recurring and regular motion of the starry sky, the sun, and the planets, as well as the regular succession of seasons, suggested to humans the idea of predictability and order. Humans have never been content to observe but have always sought to understand. The development of these ideas led them to believe in a completely ordered universe and the possibility of predicting events once considered mysterious or terrifying, such as the appearance of comets, solar eclipses, etc.

According to Plato, a divine being transported the cosmos from disorder to order and established the reign of forms and geometric proportions. The world is mathematically ordered, and it is the task of scientists to discover the laws that govern it.

Two thousand years later, two giants of human history, Galileo and Newton, made an epochal revolution by discovering the laws that govern the evolution of natural phenomena. They laid the foundations of what we today call "classical mechanics", the basis of modern science. The same laws that govern the motion of celestial bodies also govern all movements on Earth. These laws are deterministic: the future of every phenomenon occurring around us is uniquely determined by the present state. In this system, there is no room for random or arbitrary events; everything obeys a precise and predictable order.

Following the great discoveries of Galileo and Newton, the Enlightenment idea gradually took hold that humans could, at least in principle, predict and even control the course of future events. "Give me the state of the universe, and I will predict the future", Laplace declared, summarizing the unshakable faith of the era in human capabilities. This phrase embodies the spirit and enthusiasm of the time, marked by the triumph of mechanics, the discovery of heat engines, and the Industrial Revolution

Enlightenment optimism, based on determinism, reinforced the belief that humans could predict the evolution of events with increasing precision. This thought was further consolidated in the nineteenth century, thanks to extraordinary scientific achievements such as the discovery of electromagnetic waves. In this worldview, the universe was conceived as a large machine, where each part moved in a determined and predictable way. As Laplace affirmed, "to an intellect sufficiently elevated, nothing would be uncertain, and the future, as the past, would be present to its eyes".

During the twentieth century, several great discoveries cast a shadow over the optimistic picture that had dominated for more than three centuries, undermining the illusion of perfect and deterministic knowledge of the universe. We are referring, of course, to the discovery of quantum mechanics, but also to the concept of deterministic chaos. Both theories have shaken the certainties built since the Enlightenment, forcing us to abandon Laplace's ideal of a perfectly predictable universe governed by an immutable scheme. Quantum mechanics and deterministic chaos, in fact introduce an element of uncertainty and probability into our understanding of reality, limiting the possibility of knowing the future of the world around us with absolute precision.

As often happens in human history, recognizing one's limitations becomes an opportunity for new leaps forward. In this case, the boundaries imposed on human knowledge—up to the very impossibility of predicting the future—have sown the seeds of a new great scientific progress, which is now in full development. This new "technological civilization" is no longer based on certainties, determinism, and order, but is capable of embracing and exploiting concepts such as deterministic chaos, probability, and quantum entanglement. Far from being constrained by the limitations of the past, this evolution is leading us toward a richer and more nuanced understanding of the world, where uncertainty becomes a tool for progress rather than an obstacle.

1.10.1 *Deterministic chaos*

> Determinism is an illusion due to Laplace.
>
> Henri Poincaré

Deterministic chaos offers us the opportunity to overcome some deeply rooted prejudices. Readers may be perplexed by the apparent contradiction inherent in the expression "deterministic chaos", which at first glance seems an oxymoron. We are naturally inclined to think that a phenomenon is either determined or random: it cannot be both. However, this apparent contradiction stems from cognitive barriers built over centuries, where we have conceived determinism and chaos as opposing concepts. In common language, in fact, chaos is associated with disordered, uncontrolled events that seem to proceed without an apparent logic or rule: stock market crashes, revolutions, epidemics. It is the antithesis of what we expect from science—a tool to bring order and predict the evolution of natural, economic, or social phenomena. In fact, it is precisely in the notion of determinism that the scientific basis for predicting the future is contained: a system is deterministic when its future is uniquely determined by its present state.

But there is a crucial aspect we often ignore: the fact that a system is determined does not necessarily mean that we are able to predict its behavior. In recent decades, in fact, it has been discovered that relatively simple deterministic laws can generate behaviors so complex as to be completely unpredictable, to the point of being indistinguishable from chance.

In other words, even if according to classical physics the future is determined, we are not always able to know it.[4]

It may seem almost scandalous, but we do not even know for sure if our solar system is stable, or if one day the Earth will end up falling into the Sun or being expelled into space. Although we can predict the orbit of a satellite with a certain approximation and track its position moment by moment, thanks to modern computers, we are not able to obtain an exact solution to the equation that describes its trajectory. This limit, intrinsic to the generality of deterministic systems, reminds us that uncertainty does not belong only to the quantum world but is also present in the more familiar dynamics of classical physics.

But once we become aware of this strong limitation—that the behavior of most deterministic systems is in reality so complex as to appear completely chaotic and thus elude our ability to predict—what sense does the study of deterministic chaos make, and what concrete implications can it have? The answer to this question lies in the fact that, if we think carefully, we are not interested in knowing the future with absolute precision. What we are interested in is being able to make reliable predictions about the quantities we are able to measure, about the average values that influence our decisions.

Take, for example, the case of an epidemic. We are not really interested in knowing the exact number of people who will have contracted the disease at a specific time on a precise day next week. This data, besides being unpredictable, may not even be measurable with precision. What we urgently need to know is whether the number of infected people is growing, whether the epidemic has reached an oscillatory regime, or whether it is decreasing, leading to the extinction of the contagion. Similarly, we are not interested in knowing the position and velocity of every single air molecule: what we are interested in knowing are quantities such as temperature, humidity, pressure, or wind force.

In summary, the study of deterministic chaos provides us with tools to understand and predict the evolution of complex systems not in their minute details, but in their general trends and average dynamics. It is this type of information that allows us to make practical decisions and manage apparently unpredictable situations with a sufficient margin of control.

[4]More precisely, saying that we are unable to predict the future means that we cannot find an algorithm that predicts the future up to time t and whose length grows more slowly than t. Obviously, if the length of the algorithm grew at the same rate as t, this would not be a solution—i.e., not a prediction of the future—because it would be equivalent to observing the system and waiting until time t for the future to unfold.

It is therefore fortunate that, as has been discovered, most phenomena follow the laws of deterministic chaos, while perfectly ordered systems—for which we are able to find an exact solution—are rare, true exceptions. An ordered system, in fact, is structurally unstable: a small perturbation is enough to deviate it drastically from its initial behavior, like a stone balanced on top of a mountain, which with a minimal movement falls into a completely different state. On the contrary, a chaotic system is structurally stable: even if perturbed, it remains chaotic, maintaining a certain coherence in its disorder.

And it is precisely this stability of chaos that allows us to make predictions and understand which parameters to act on to influence the evolution of phenomena. In the already mentioned case of an epidemic, we will never be able to calculate the exact number of infected people at a given moment, but we can develop methods to identify the critical parameters, limits, and conditions that determine whether the epidemic will grow or become extinct. In this way, it is possible to identify the most effective actions to contain or combat it. If the systems around us—not only natural ones but also economic and social ones—were perfectly ordered, it would be a real disaster. We could not make reliable predictions, and every small change in parameters would lead to drastically different consequences, making it impossible to control events. In such a rigid world, our ability to manage and understand reality would be severely limited.

The study of deterministic chaos in the world of classical physics originated precisely from the recognition of a fundamental limitation: the impossibility of predicting the future with exactitude. One of the greatest scientists in history, James Clerk Maxwell, founder of electromagnetism, wrote with extraordinary intuition as early as the second half of the nineteenth century: "The true logic of this world is the calculus of probabilities". And surprisingly, he was referring to the world of classical mechanics. Quantum mechanics not only confirmed and reinforced this concept but also imposed even deeper and more unexpected limits, with consequences that even its own pioneers could not have imagined. These new restrictions have revealed an even more complex and uncertain universe, testing our ability to understand and predict.

Chapter 2

Beyond Certainties: Logic and Power of Quantum Computing

It is not worthy of excellent
men to lose hours like slaves
and toil over calculations that
could be entrusted to anyone if
machines were used.

Gottfried Wilhelm von Leibniz

2.1 Quantum Logic

Isidor Isaac Rabi, Nobel Prize winner in Physics, once said that "science is an adventure of the entire human race to learn to live and, perhaps, to love the universe in which it lives." In this adventure, extraordinary figures emerge from time to time—pioneers who push humanity beyond the boundaries of the known, anticipating developments that will take shape only many years later. Among these giants is Richard Feynman (1918–1988), one of the most brilliant physicists of the twentieth century, who in 1981 put forward a revolutionary idea: the construction of a quantum computer.

His intuition was as simple as it was bold: quantum systems behave in an incredibly complex way, so why try to simulate them with classical computers, which follow completely different logics? Would it not make more sense to use a machine that obeys the same rules as the systems it must simulate? Precisely a quantum computer. At that time, most physicists considered the idea little more than an intellectual game, an unrealistic dream. (One of the authors of this book still remembers the smiles exchanged among experts at an international conference in the 1980s, when someone dared to mention Feynman's idea.) Today, what once seemed pure science fiction has become one of the greatest scientific and technological challenges of our time.

Making predictions about the future is always risky: you can end up spectacularly off track. One of the most wrong predictions in history is attributed to Thomas Watson, president of IBM, who in 1943 is said to have stated that the world would not need more than five computers. And we are not talking about any visionary, but the leader of the largest technology company of the time. Yet, reality has continued to contradict even the greatest experts. In 1977, Ken Olsen, founder of Digital Equipment Corporation, declared that he could not imagine why anyone would ever want a computer in their home. Today, that once improbable future is our present: there are more than a billion computers in the world, not counting those integrated into everyday objects such as mobile phones, watches and cars. Predictions, no matter how formulated by experts, rarely keep pace with the incredible speed at which technology transforms our world.

We are at the dawn of a new conceptual technological leap in the field of computers and their possible future use: the advent of quantum computers. In the following paragraphs, we will delve into their operation. At first glance, we might be tempted to think that these computers are too complex to enter our homes in the future. But in light of what we have already seen with classical computers, we prefer to avoid hasty predictions: history has already surprised us once, and it could do so again.

But what really is a quantum computer? What is the logic behind this extraordinary machine? And why does it have the potential to be enormously more efficient than classical computers? The basic principle remains the same: a computer, whether classical or quantum, takes input information and processes a result (output) through an algorithm. Whether it is solving complex mathematical equations, designing new materials and drugs, optimizing industrial processes, forecasting the weather, or simply browsing the Internet or playing a video game, the core of the process is always the processing of information. However, quantum computers promise to do all this in ways that today seem almost science fiction.

Since ancient times, humans have sought to simplify calculations with the help of machines. One of the first tools we know of is the abacus, introduced in China around two thousand years before Christ. But the true precursor of modern computers arrived in the nineteenth century, thanks to the ingenuity of the English mathematician Charles Babbage. His project, a programmable computer, laid the foundations for modern computing.

Working alongside him was Ada Lovelace, daughter of the famous poet Lord Byron, who today is remembered as the first computer programmer.

It is curious to think that in 1878, the British Association for the Advancement of Science—at the time one of the most prestigious bodies in the scientific community—recommended to the British government not to fund Babbage's machine, considering it "not worthy of support". An evaluation error that, in light of subsequent technological developments, now seems almost ironic.

The operation of a modern computer, no matter how complex it may seem, is surprisingly simple in its basic principle. At its root is the binary number system, formalized by the German mathematician and philosopher Gottfried Wilhelm Leibniz. In this system, the fundamental unit of information is the bit, which can only take two states: 0 or 1. A classical computer is a tool that accepts information in binary form, processes it, and always returns the result in binary form.

We can imagine the inside of a computer as a network of wires and logic gates. Each wire carries a bit, which can be 0 or 1, while the logic gates perform mathematical or decisional operations on the bits, transforming them and generating new results. This binary language is perfectly suited to electronic machines: to represent any number, all you need is a series of 0s and 1s. For example, 0 can indicate that current flows through a wire, while 1 indicates that there is no current. Thanks to this simplicity, computers are able to process enormous amounts of information at impressive speeds.

2.2 The Turing Machine

The word "algorithm" derives from the name of the Persian mathematician and astronomer Al-Khwārizmī[1] and denotes a procedure (a finite sequence of instructions) for solving a specific problem. This concept, fundamental to computer science, found its precise mathematical formulation thanks to the Turing machine, introduced in 1936 by the brilliant English mathematician Alan Turing.

The Turing machine represents a theoretical model that captures the essential elements of any modern computer: memory to store data, a control

[1] Al-Khwārizmī, who lived between 780 and 850 CE, worked in Baghdad at the Bayt al-Hikma (House of Wisdom), an extraordinary institution founded by Caliph Al-Ma'mūn. This center was the vibrant heart of scientific and cultural research of the era, where Greek, Indian, and Persian texts were translated and studied. The term "algorithm" derives from the Latinization of his name, "Algoritmi" and spread from the Latin translation of his treatise on arithmetic.

unit that directs operations, and a read/write unit to manipulate information. This simple yet powerful concept laid the foundations for understanding what a computer is capable of doing, making possible the entire digital universe we know today.

The fundamental elements of a Turing machine are represented in Figure 2.1 and offer a simple but powerful vision of the operation of this theoretical machine. The underlying idea is that the machine performs calculations in a way similar to a "human computer". Imagine a human being who, although able to store only a limited amount of information in their brain, has access to a quantity of paper (ideally infinite) to write down and consult data during read and write operations. Similarly, the Turing machine consists of three fundamental elements that work in harmony to perform calculations:

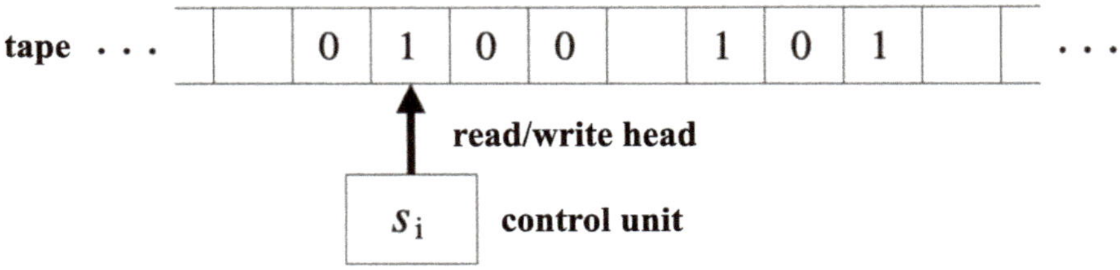

Fig. 2.1. Schematic drawing of the Turing machine, with the tape, the read/write head, and the control unit.

(1) **The tape:** represents the "paper" of the Turing machine, an ideal tape that extends infinitely and is divided into cells. Each cell can contain a letter from an alphabet (for example, in a binary alphabet, 0 or 1) or remain empty. Except for a finite number of cells containing data, the rest of the tape is empty.

(2) **The control unit:** the "brain" of the machine, which can be in one of a finite number of states. One of these states is the special halt state (H). When the control unit enters state H, the calculation ends. Until then, the control unit supervises the process, deciding what to do based on the instructions and information it receives from the tape.

(3) **The read/write head:** at any moment, this head is pointing to a single cell of the tape. Initially, the data of the problem to be solved is encoded on the tape. The calculation then proceeds cyclically

as follows: depending on the internal state s_i of the control unit and the content of the cell under the head, it can read, write, or erase the symbol in the cell. After performing the operation, the head moves one position to the right or left on the tape, or the calculation ends if the control unit has reached the halt state (H).

These three elements together form an extremely versatile system, capable of performing a wide range of calculations, making the Turing machine one of the most influential theoretical foundations for modern computer science.

The operation of a Turing machine is guided by a finite set of rules, each of which, based on the internal state of the control unit and the letter that the head reads on the tape, determines what the machine itself must do. In practice, given an initial state, the rules establish which actions the head must perform (read, write, erase, or move) and in which new state the control unit must be after each operation. It is this mechanism of instructions that allows the machine to perform calculations, solve problems, and ultimately simulate any computational process that follows defined rules.

Despite its simple and minimal structure, the Turing machine contains extraordinary power: it is capable of tackling and solving problems of extreme complexity. Surprisingly, it can perform any operation executable on a modern computer. More precisely, if a problem has an algorithmic solution, the Turing machine can compute it. In other words, any computational process, regardless of its complexity, can be performed by a Turing machine.

Alan Turing's work took shape within a lively debate that dominated the era, centered on a crucial question: for which classes of problems is it possible to find a solving algorithm? In 1928, the mathematicians David Hilbert and Wilhelm Ackermann formulated a question known as the Entscheidungsproblem (decision problem), asking whether there existed an algorithm capable of determining, for every mathematical statement, whether it was true or false, thus returning a clear "yes" or "no" as an answer.

Hilbert was mistakenly convinced that the answer to this question was positive, and hoped for a universal solution to mathematical reasoning. However, in 1931, the Austrian mathematician and philosopher Kurt Gödel dismantled this vision with his famous incompleteness theorem. Gödel proved that in any logical system "sufficiently powerful" to formalize arithmetic, there exist undecidable propositions, i.e., propositions whose truth or falsity cannot be proven within the system itself.[2]

Logical contradictions within a formal system can be understood through simple examples called paradoxes. One of the most famous is the sentence: "This sentence is false". Its grammatical structure is perfectly correct, but we cannot answer the question: "Is the statement true or false?" If it were true, then it would mean that it is false, but if it were false, then it should be true, generating an infinite cycle of contradictions.

Another example is Buridan's paradox, devised by the French philosopher Jean Buridan in the fourteenth century. Imagine two statements: Socrates says: "Plato tells a lie" and Plato asserts: "Socrates tells the truth". Taken individually, each statement is free of logical problems, but together they lead to a paradox. If we assume that Socrates tells the truth, then Plato lies, and thus Socrates is telling a lie, which contradicts the initial assumption. Conversely, if we start with the idea that Socrates tells a lie, Plato should be telling the truth, but this would mean that Socrates is telling the truth, bringing us back to a contradiction. These examples illustrate how some propositions, even when constructed in a perfectly logical way, cannot be definitively solved, revealing the intrinsic limits of logic.

Gödel's theorem implies that not all mathematical statements can be solved with an algorithm, and this stimulated Turing's thinking about what a computer could really do. Gödel's theorem also places limits on the capabilities of a Turing machine, and thus of any computer, classical or quantum. There exist undecidable problems, i.e., problems that cannot be solved by any algorithm—and this limit applies equally to classical and quantum computers. In other words, a quantum computer cannot solve problems that are, by their nature, unsolvable by a classical computer. However, quantum computers offer a significant advantage: they can solve some problems much faster than a classical computer, using fewer resources in terms of memory, computation time, and energy.

[2]It should be noted that Presburger arithmetic does not contain undecidable propositions. However, it is not "sufficiently powerful", as it is constructed for natural numbers (i.e., the set of numbers 0, 1, 2, 3...) with only the operation of addition. Undecidable propositions, though, appear in a more powerful arithmetic—Peano arithmetic—which also includes multiplication.

2.3 Quantum Parallelism

> Its fire followed every spark;
> and they were so many that
> their number stretches to more
> than the doubling of the
> chessboard.
>
> Dante Alighieri, *Paradise, Canto XXVIII, vv. 91–93*

As we explained in the first chapter, the classical bit can only be in two states, 0 and 1, but can never be in both simultaneously: it is either in state 0 or state 1. Unlike the classical bit, which corresponds only to the North Pole (0) or the South Pole (1), the quantum bit (or qubit) can instead be in a superposition of the 0 and 1 states, and we visualized this state as any point on the surface of a globe.

This extraordinary property of qubits is the basis of quantum parallelism, from which the great potential of quantum computing derives. But what does this exactly mean? The concept of parallel computing already exists in classical computers, where, to accelerate the execution of a complex problem, the work is divided among several processors. More processors work in parallel to obtain the result in a shorter time. To illustrate the concept of quantum parallelism, take the famous traveling salesman problem: a salesman must visit a certain number of clients, each in a different city, and then return home. His goal is to visit all clients in the order that minimizes the total distance traveled. This means having to consider every possible order of visits, calculating which is the shortest path. Imagine the salesman starts in Milan and must visit other Italian cities, Piacenza, Genoa, Parma, Cremona, Mantua, and Bergamo, before returning to Milan. One possibility is to follow the order we just described, but he could also go to Bergamo first, then Mantua, then Cremona, and so on. With six cities to visit, there are 720 possible routes. If the number of cities were 10, the combinations would exceed one million. With 15 cities, the possible routes would be more than a trillion! Each route represents an input for which we must calculate the total distance traveled. If we assigned each calculation to a different processor, we would have a form of parallel computing. However, a problem arises here: we cannot have a trillion processors. Classical computers are therefore limited by the available physical resources. In quantum computing, instead, thanks to state superposition, qubits can simultaneously explore multiple paths in the space of possibilities, enabling an incredibly more powerful form of parallelism and opening new perspectives for solving complex problems.

In essence, quantum mechanics adds a completely special dimension to the concept of parallel computing that cannot be replicated in classical computers. To understand this, let's start with a simple example: imagine we have a single classical bit with which we can only perform trivial operations. For example, if the bit is represented by a coin, where heads and tails stand for states 0 and 1, respectively, the operation we can perform is flipping the coin, i.e., if we start with heads as input, we get tails as output, and vice versa, starting with tails gives heads. However, we cannot simultaneously perform the calculation (flip the coin) for both possible input states, 0 and 1.

In the quantum world, the rules change radically. Unlike a classical bit, a qubit can be in a superposition of the 0 and 1 states simultaneously. Therefore, it is possible to perform a given calculation simultaneously with both state 0 and state 1 as input. This is quantum parallelism, whose importance we can grasp when we think about what happens when we increase the number of qubits in which we encode an input through a sequence of 0s and 1s. With two qubits, the possible input states with which we can perform calculations simultaneously become 4 (00, 01, 10, 11), with three qubits they become 8 (000, 001, 010, 011, 100, 101, 110, 111), and so on.

The growth of input states on which we can perform calculations in parallel is very rapid with the number of qubits. More precisely, it is exponential. This is a growth law we unfortunately got to know well with the coronavirus. In the case of an infectious disease with a basic reproduction number R_0 equal to 2, each infected person, in the absence of containment measures, would infect an average of two other people. This type of exponential growth may initially seem manageable, but it can quickly spiral out of control. For example, if the average virus transmission time is one week, starting from a single case, we would reach about a thousand cases in two months, a million in four months, and a billion in six months. Naturally, such exponential growth cannot continue indefinitely, as there are physical limits, such as the world population, which in 2025 is about eight billion people. Similarly, the power of quantum computers increases exponentially with the number of qubits, allowing calculations to be performed on a number of simultaneous states inaccessible to classical parallel computers.

To understand the incredible impact of exponential growth in quantum parallelism, we can think of a less dramatic example than the previous one: that of the pharaoh and the chess master. According to legend, the master who taught the pharaoh to play chess asked for a reward of one grain of wheat on the first square of the chessboard, two on the second, four on the third, and so on, doubling the number of grains on each square. The pharaoh, convinced that the request was modest, agreed, but soon realized that not even the entire harvest of his kingdom for many years would be sufficient. In fact, to satisfy this request, he would have had to deliver ten billion billion grains of wheat just for the last square.

Similarly, a quantum computer with 64 qubits could simultaneously perform a calculation for a number of input states equal to the number of grains of wheat the pharaoh would have had to deliver to his chess master. Such a computer, leveraging quantum parallelism, would therefore have potential inaccessible to any classical computer. It could solve in seconds problems that today would require the most powerful supercomputers more time than the age of the universe. It is precisely this exponential growth that gives quantum computing its revolutionary potential, capable of tackling problems that were previously considered unsolvable.

2.4 Elementary Logic Gates

How do calculations actually take place inside a computer? To answer this, let's imagine for a moment going back to school, when the teacher taught us to add two numbers. We proceeded column by column, starting from the right: units with units, tens with tens, and so on, handling any carries. For example, if we add 8 and 4, we get 12. We write 2 in the units column and "carry over" the 1 to the tens column. Well, the computer does exactly the same type of operation, except that instead of decimal numbers, it uses binary numbers. In this representation, addition follows specific but equally simple rules. For example:

$0 + 0 = 0$,
$0 + 1 = 1$,
$1 + 0 = 1$,
$1 + 1 = 0$ with a carry of 1.

To execute an algorithm, such as an addition operation, a classical computer breaks down the process into a series of elementary steps called logic gates. Among the simplest are the NOT, COPY, and AND gates. The NOT gate is like a switch: if it receives 0, it transforms it into 1, and vice versa. The COPY gate, on the other hand, creates an exact copy of a bit without altering it. We can imagine a bit as the result of a flipped coin: heads or tails. Well, the NOT gate is like flipping the coin, while the COPY gate prepares another coin with the same result as the first. Then there is the AND gate, which works on two bits simultaneously: if both "coins" show heads (i.e., 1), then it returns 1; in all other cases, the output will be 0. These simple logical building blocks form the basis of any calculation inside a computer.

How can we implement these logic gates inside a computer that uses electrical signals? The principle is simple: imagine the logic gates as boxes with wires coming in and out. The state of a bit, i.e., 0 or 1, corresponds to whether current flows through a wire (1) or not (0). Take, for example, an AND gate. Suppose two wires, A and B, enter the box, representing the states of the physical input bits, and one wire, C, exits, representing the output bit. The AND gate works such that wire C will carry current only if wires A and B are both carrying current. In other words, the condition for the output to be 1 is that both inputs are 1. This type of logical operation is then electronically realized through circuits, where wires A and B act as inputs and C acts as output. With a combination of electronic components such as transistors, these circuits physically make it possible to implement logic gates inside a real computer.

But how is it possible, starting from these simple logic gates, to perform complex calculations? Here lies the real magic: it can be proven that with only three basic operations—NOT, COPY, and AND—we can build any type of calculation, no matter how complicated it may seem.[3] It is extraordinary to think that by combining these three simple gates, we are able to realize all imaginable logical functions. In practice, this means that we can execute any algorithm that solves any problem, simply by orchestrating these elementary operations. It's a bit like building an entire castle with just a few types of bricks: the key lies in how you put them together!

The beating heart of every modern computer is therefore a small but extremely powerful component: the transistor. It is thanks to this small electronic device that logic gates become a reality and that computers can perform any operation. In today's computers, a single microprocessor can contain over a billion transistors, which work in unison to run

[3] In fact, a single type of gate—such as the NAND gate—is sufficient and can be realized using transistors. The NAND gate is simply the negation of the AND gate: it outputs 1 whenever the AND gate would output 0, and vice versa.

every calculation. The invention of the transistor in the late 1940s marked the beginning of a true technological revolution, giving rise to an unstoppable growth of the electronics industry. Without the transistor, the digital world we know would not exist!

The ability to break down any algorithm into elementary logic gates is one of the cornerstones of modern computer science. This allows us to use the same hardware—the same electronic circuits—to solve any type of problem, without having to change the underlying infrastructure. In other words, once we have transistors, there is no need to add new electronic components to perform calculations. However, if, for example, we increase the number of digits of the numbers to be added, we must also increase the number of basic operations such as NOT, COPY, and AND, and consequently the time required to complete the calculation will also increase. In addition to time, memory also plays a fundamental role: as numbers become larger, we need more space in the computer's memory, i.e., more cells, each identified by a unique "address". Here, input data and the final result are stored, ready to be read or printed. In recent decades, all these difficulties have been successfully overcome thanks to innovations such as integrated circuits, which contain millions of transistors and have allowed classical computers to become increasingly powerful and capable of tackling problems of extraordinary complexity.

Just like in a classical computer, every calculation in a quantum computer can be broken down into a sequence of elementary logic gates. However, these gates do not act on simple bits, but on qubits, and can operate on one or two qubits at a time. The real revolution lies in the nature of quantum logic gates: while a single-bit classical gate can only switch the state between 0 and 1, a quantum logic gate has the power to bring a qubit into a superposition of 0 and 1, i.e., in the globe image, the quantum logic gate can move the qubit from one point to another on the surface of the sphere, opening up an entire universe of possibilities. This flexibility is what makes quantum computing so extraordinary and powerful.

The fundamental quantum logic gate involving two qubits is called CNOT, or "controlled-NOT". Its operation is simple but powerful: the first qubit, called the control qubit, remains unchanged. It is the second qubit, the target qubit, that can undergo a transformation: if the control qubit is in state 1, the CNOT "flips" it, switching from 0 to 1 or 1 to 0, just like a NOT gate. If the control qubit is in state 0, the gate does nothing, leaving everything unchanged. But the true characteristic of the CNOT gate is that even if we start with two separate, non-entangled qubits, its effect is generally to create entanglement: in other words, it puts the two qubits into a state where they can no longer be described independently of each other. This ability to generate entanglement, one of the most fascinating characteristics of quantum physics, is what makes CNOT one of the fundamental components for the power of quantum computers.

It is interesting to note how quantum logic gates, both single-qubit and CNOT, allow us to leverage two of the most incredible characteristics of quantum mechanics: superposition and entanglement. It is precisely these properties, so difficult to reconcile with our intuition, that open the door to the extraordinary power of quantum computing. On the one hand, they seem almost counterintuitive, but on the other, they represent the key resources that enable quantum parallelism, the real secret of the computational power of a quantum computer. And here lies the beauty: since every quantum calculation, as we mentioned earlier, can be broken down into these simple elementary gates, the quantum machine becomes capable of tackling problems that would seem impossible for classical computers.

So everything is solved? Unfortunately not, the reality is more complex. So far, we have neglected a crucial aspect of quantum mechanics: to obtain the output, i.e., the result of the calculation, we must interrogate the computer through a measurement. And this is where a first problem arises. As we know, the measurement process in quantum mechanics is inherently probabilistic, and the probabilities of possible outputs are governed by the postulates of quantum physics. This means that even though the quantum computer can perform an infinite number of calculations in parallel for a large number of input states, in the end we still have to perform a measurement to know the result. And the measurement for each qubit returns only a classical bit, 0 or 1, with a determined probability. In other words, the quantum computer may have performed an impressive number of calculations, but in the end, we only have access to the measurement result, a single classical bit for each qubit involved. This makes it seem like the enormous advantage of quantum parallelism dissipates, leaving us with a single answer, and what's more, not always the same one, given the probabilistic nature of the measurement.

Is there a way out? Can we really exploit all the information contained in a quantum computer? Even though this obstacle seems insurmountable, the answer is yes! For years, scientists and researchers have been working tirelessly to solve this challenge, developing clever quantum algorithms. These algorithms are designed to leverage the power of quantum parallelism, combining and superimposing the results of parallel calculations such that, at the time of measurement, the qubits return exactly the answer we are looking for. It is not a simple task: often, due to the probabilistic result of the measurement, it is necessary to repeat the calculation several times before arriving at the desired solution. However, the potential advantage of quantum parallelism is enormous and, as we will see shortly, can lead to extraordinary results.

2.5 Quantum Algorithms

2.5.1 *Integer factorization*

"The problem of distinguishing prime numbers from composite numbers, and of resolving composite numbers into their prime factors, is one of the most important and useful in all arithmetic. The dignity of science seems to require that every aid to the solution of such a celebrated and elegant problem should be zealously cultivated". So wrote the great mathematician Carl Friedrich Gauss in 1801 in his Disquisitiones Arithmeticae. But what use can prime numbers have? Certainly, in Gauss's time, they had no practical use, and even we, when we learned about their existence from our teacher, always thought they were a mathematical curiosity, of no interest for our real lives. We remember that prime numbers are all numbers divisible only by 1 or themselves. So, 2, 3, 5, 7, 11, 13, 17, 23... are prime numbers. A prime number, i.e., cannot be obtained by multiplying two integers other than 1.

They may seem like a theoretical oddity, but over time, it has been discovered that prime numbers hide secrets of enormous relevance, which today have an extraordinary impact on the modern world. Why? The answer lies in a surprising property. Consider two prime numbers p and q. Calculating their product $N = p * q$ is easy. However, if we lose track of the two numbers p and q and are left only with N, it becomes extremely difficult to trace back to the original prime factors p and q. Let's take an example: multiplying 97 and 109 gives us 10,573, an operation we can easily perform. But if we are only given the number 10,573, finding that its prime factors are exactly 97 and 109 is not so straightforward. And the difficulty grows exponentially as the size of the numbers increases. This is the core of the problem: the more digits a number has (or the more bits are needed to represent it), the more complex it becomes to break it down into its prime factors. In fact, the best algorithm available today for factorizing large numbers requires a number of operations (or a number of logic gates) that grows exponentially with the number of digits.

It's like the anecdote of the pharaoh and the grains of wheat: the problem seems simple at first, but becomes unstoppably difficult. While finding that the prime factors of 15 are 3 and 5 is child's play, factorizing a 250-digit decimal number (i.e., 829 bits) was only achieved in 2020, using supercomputers with a large number of processors, resources equivalent to the use of a single processor (at the current state of the art) for 2700 years. If we tried to factorize a 300-digit number, it would take more than a million years with current technologies. In practice, it becomes an impossible problem to solve, even with the computing resources of the entire human race.

The operation of cryptographic codes, which allow us to transmit information confidentially, is based precisely on an apparently insurmountable challenge: the difficulty, which in practice translates to impossibility, of factorizing large integers. And we are not just talking about military secrets, industrial data, or diplomatic messages between governments. In the age of the Internet, cryptography is at the center of our daily lives: every time we buy something online, book a vacation, or send a personal message, we rely on these systems. None of us wants our credit card data to end up in the wrong hands, such as criminal organizations, or our private information to be disclosed.

This is why computers encode (encrypt) our messages, making them unreadable to anyone trying to intercept them. Only once they reach their destination are these messages decoded (decrypted) and made understandable again. This encoding and decoding system is now integrated into the main communication protocols and the most widely used operating systems. Its security depends precisely on the fact that the factorization of large prime numbers, necessary to decrypt a message, is simply unfeasible with current computers. Even with the most advanced technology, the time required to solve this problem would be so long as to make it impossible.

What is the connection between all this and quantum mechanics, or a quantum computer? So far, it might seem intuitive that a quantum computer is useful for studying quantum systems, but it is much less obvious to understand how it can be incredibly more efficient than a classical computer at solving problems that have apparently nothing to do with quantum physics. For many years, after Richard Feynman's original proposal, the quantum computer was considered a fascinating idea but destined to remain only theory. This changed radically in 1994, when a mathematician named Peter Shor changed the game. Shor published an algorithm, specifically designed to solve a problem far from quantum physics: integer factorization.

His algorithm, executable on a quantum computer, is able to find the prime factors of a number extremely efficiently, a challenge that classical computers are unable to tackle. This discovery transformed the perception of the potential of quantum computing, and for his work, Peter Shor received the prestigious Breakthrough Prize in Physics in 2022, along with pioneers of quantum information such as Charles Bennett, Gilles Brassard, and David Deutsch.[4]

It is not necessary to explain in detail how Shor's algorithm works, but the basic idea is as follows: Shor managed to transform the factorization problem into that of finding the "period" r of a specific periodic function—a function is periodic if it repeats itself after a certain interval r. The efficiency of the algorithm stems from the fact that it leverages quantum parallelism, allowing the quantum computer to evaluate this function simultaneously for a large number of inputs. The real stroke of genius lies in the ability to combine all these parallel calculations such that, in the end, by measuring the state of the qubits, the period of the function is obtained. It can be proven that once the period of the function is found, the problem of factorizing an integer can be solved quickly with a normal classical computer.

Shor's algorithm is so powerful that a quantum computer with a few thousand qubits could break the cryptographic codes currently in use, which are based precisely on the difficulty of factorization. It is true that today we are still far from this scenario: in the laboratory, we have only been able to factorize small numbers. But we can imagine a not-too-distant future where the rapid advancement of quantum technologies will seriously threaten the security of encrypted information. Fortunately, quantum mechanics already holds a solution: as we will see later, it offers an alternative and inherently secure method for transmitting sensitive information. It is almost as if quantum physics itself, which could undermine our security, simultaneously offers us the key to protecting it.

2.5.2 *Searching a database*

We live in the era of big data, a time when enormous flows of information are collected by organizations, companies, and governments every second.

[4]The Breakthrough Prize was established by Sergey Brin, Priscilla Chan, Mark Zuckerberg, Yuri and Julia Milner, and Anne Wojcicki. The prize amount is 3 million US dollars, approximately three times the Nobel Prize.

These data comes from an impressive variety of sources: commercial transactions, smart devices, industrial machinery, videos, social media, and much more. They are stored and analyzed to extract useful information that guides business decisions, shapes strategies, reduces costs, accelerates processes, and fuels innovation. In essence, they have become a valuable resource for those who know how to use them to optimize offerings and develop new products.

However, managing this immense amount of data is not at all simple. The greatest challenge is knowing how to filter and select the truly useful information—information that meets specific needs and objectives. It is a crucial and priority problem for every company or organization navigating this sea of data. To give an idea of the scale of the phenomenon, consider that almost one zettabyte of data is generated every day—the equivalent of a 1 followed by 21 zeros in bytes, where 1 byte is made up of 8 bits. Faced with this staggering amount of information, the question naturally arises: how can we solve this problem?

In 1996, the Indian computer scientist Lov Kumar Grover invented a brilliant algorithm capable of solving a problem that is both complex and crucial: finding a specific element within an unstructured list (or database). Imagine, for example, having to search for a contact in a phone book. If you know the person's name, it is easy to find their number because the names are sorted alphabetically. But if you only have the number and need to find the name, the situation changes drastically: you have no choice but to flip through page after page, scrolling through each number until you find the right one. Sometimes you will be lucky and find it in the first few pages, but other times you will have to go almost to the end. On average, the search requires consulting half of the list, a process that, if the database is large, can take a considerable amount of time. However, Grover found a solution that, leveraging the power of quantum computing, makes this search enormously more efficient.

This example may seem like a special case, but searching for an element in a database is an incredibly common and relevant problem. After Shor's algorithm, Grover's algorithm is considered one of the most important in the world of quantum computing. Its strength? It allows finding the desired element with impressive speed. Thanks to quantum parallelism, the search time is no longer proportional to the number of elements in the database, but to its square root. The advantage is enormous. Returning to our example, consider a phone book with a million numbers. If each search takes one second, without Grover's algorithm, the worst-case scenario would require a million seconds, or about 12 days. With Grover's algorithm, that time is drastically reduced to just 1000 seconds, a little over a quarter of an hour! And the larger the database, the greater the gain: for a billion numbers, Grover would reduce the time from 31 years to just 8 hours. An incredible saving that demonstrates how powerful and useful this algorithm is.

2.5.3 *Quantum algorithms and measurements*

Later, we will explore the wide range of complex problems that a quantum computer can tackle. For now, let's focus on a fundamental aspect of every quantum algorithm: at the end of the calculations, the quantum computer must be interrogated through a measurement process, from which 0 or 1 is obtained for each qubit, with probabilities determined by the laws of quantum physics. From this sequence of 0s and 1s, we want to obtain a clear answer to the problem we are interested in.

For example, in Shor's algorithm, such a sequence of bits represents the binary encoding of the period of a function, from which we derive the prime factors of a number. In material simulation, the sequence of 0s and 1s could reveal the value of crucial properties such as the electrical or thermal conductivity of the material under study. The speed with which a quantum computer can provide answers depends not only on the nature of the problem but also on our ability to develop efficient quantum algorithms.

The underlying difficulty is that in quantum mechanics, measurement is inherently probabilistic, which means that an algorithm will not always guarantee the correct result on the first try. However, this uncertainty is not a cause for concern. If the probability of success is sufficiently high, we can reach the correct solution with a relatively small number of repetitions of the algorithm. In fact, just like in any analysis based on probability calculations, there is always the possibility that the algorithm may fail repeatedly. But the beauty of quantum computing lies in the fact that as the number of repetitions increases, the probability of continuing to fail decreases exponentially. For example, if after a hundred attempts the probability of not having yet obtained the correct result is less than one in a billion, we can consider success practically guaranteed within that number of attempts.

2.6 Schrödinger's Dilemma: Decoherence and the Elusive Cat

The advent of quantum algorithms capable of tackling problems previously unexplored by classical computers has given a strong impetus to the development of quantum computers. Since crossing the threshold of the new millennium, the global academic landscape has seen an exponential growth in laboratories dedicated to exploring quantum computers and related technologies. Today, there is no physics department that does not offer at least one course on quantum computers—an area that until the early 2000s was an almost unexplored territory. This rapid evolution invites us to reflect on the dynamics of our educational system. In an era where knowledge can quickly become obsolete, it is no longer sufficient to flood young people with notions. Instead, it is fundamental to refine their critical thinking skills, equipping them with the flexibility necessary to navigate and solve the new challenges emerging from the continuous advancement of scientific knowledge. The statement by Plutarch thus becomes increasingly relevant: "The young are not vessels to be filled but fires to be kindled."

Given the incredible potential of quantum computers and the vast resources invested in realizing them, it is legitimate to ask why we do not yet see them—if not on our desks, at least in large computing centers, ready to revolutionize our approach to knowledge and technology, solving problems previously inaccessible to classical computers. Instead, as we will discover later, we are still in the prototype development phase, although these are evolving rapidly and becoming increasingly sophisticated. The main obstacle, and thus the greatest challenge, remains the phenomenon of decoherence, which undermines the stability and effectiveness of the quantum computer. Overcoming this obstacle will undoubtedly be a difficult task, requiring enormous efforts and continuous innovation.

Decoherence occurs when a quantum system comes into uncontrolled contact with the external environment, altering the state of the quantum computer and causing the loss of the information it stores and processes. Imagine the scene of a swan gliding elegantly on the calm waters of a lake, leaving behind a trail of soft and ordered waves. Now, imagine that sudden external agents such as rain, wind, or other disturbances disrupt the surface of the water, creating chaotic ripples that damage and disperse those delicate waves. Similarly, the quantum computer, in its extreme sensitivity, requires protection from the surrounding environment to maintain its integrity. Think of a qubit as a coin standing on its edge: the slightest interference, even the most gentle and imperceptible, can make it fall to one side or the other, thus determining the result of heads or tails.

When we venture into executing an algorithm, our task becomes similar to that of a conductor tuning the harmony of a group of musicians: we must manipulate the quantum computer with great precision, controlling the state of each qubit and turning on or off the interactions between the qubits to realize gates such as CNOT. This requires an almost hermetic isolation of the computer from the external environment to shield it from destructive influences such as electric and magnetic fields and radiation emitted by nearby hot bodies. The goal is to avoid the phenomenon of decoherence at all costs—that insidious interaction of qubits with the outside world that could undo the delicate superpositions of quantum states, thus destroying the advantages of the quantum computer. In this meticulous ballet of precision, we must protect and preserve the integrity of quantum operations, ensuring that our instrument is immune to the chaotic forces that seek to disturb it.

This is the moment when a crucial question might arise in the reader's mind. We have explored how microscopic systems obey the laws of quantum mechanics, which give rise to phenomena such as state superposition and entanglement. But how small must an object be for us to witness such wonders of nature? In everyday life, these phenomena remain invisible, hidden from the human eye. We therefore ask: is there a dimensional boundary beyond which the veil over the quantum world is lifted, making way for the more familiar laws of classical physics? And how can we determine when an object is small enough to dance to the bizarre rhythm of quantum mechanics? These are questions at the boundary between the visible and the invisible, challenges that stimulate our curiosity.

This question is masterfully addressed in Schrödinger's cat paradox, a thought experiment devised in 1935 by Erwin Schrödinger, one of the founding fathers of quantum mechanics. Imagine facing a sealed box containing a radioactive source, a radiation detector, a hammer, a vial of poison, and a cat. The radioactive source is a system (atom) that can be in two levels: the ground state 0 and the excited state 1. If the atom is in state 1, it can decay to state 0 with a certain probability by emitting light (a photon), which activates the detector and triggers a chain reaction: the hammer smashes the vial, releasing the poison and killing the cat.

Assume that initially, the state of the combined atom-cat system is: "atom in state 1, cat alive". Over time, the atom may or may not decay to state 0, and the cat's fate depends on this event. If the atom has decayed, the cat is dead; if not, the cat is alive. At first glance, there is nothing surprising—probabilities are in fact widely used in everyday life. For example, if we flip a coin, before looking at the result, we can say, if the coin is not rigged, that it has an equal probability of being in the "heads state" or the "tails state".

What, then, is so paradoxical about Schrödinger's cat? The point is that, according to quantum mechanics, until we observe the inside of the box, the atom and the cat are simultaneously in a superposition of "atom in state 1, cat alive" and "atom in state 0, cat dead". We are therefore faced with a paradoxical aspect for our classical intuition: the superposition of the cat's alive and dead states. This coexistence is different from the classical superposition of two primary colors that generate a new color; for example, blue and yellow mixing to create green. Unlike colors that blend harmoniously, the superposition resulting from the cat's alive and dead states defies all classical logic and cannot be observed directly, making it profoundly paradoxical. Unlike secondary colors such as green, the "secondary state" of cat alive/cat dead is never observed. This is the beauty and enigma of Schrödinger's cat, which forces us to reconsider what we think we know about reality itself.

To escape this absurd situation, the phenomenon of decoherence comes to the rescue. The atom-cat system is never perfectly isolated because it is always in some way in contact with the surrounding environment. This interaction with the environment acts, in fact, as a measuring device, revealing whether the cat is alive or dead. The presence of a conscious observer, such as a human being, is not necessary to perform the measurement. The environment itself, by interacting with the cat, disturbs its state, just as we would if we observed it directly. According to the laws of quantum mechanics, following this interaction, the cat "collapses" into a definite state: alive or dead. In other words, the interaction with the environment destroys the delicate entangled state that had formed between the atom and the cat, thus breaking the quantum uncertainty.

Schrödinger's cat paradox undoubtedly captures our imagination, but what would happen if instead of a cat, we took a much smaller living being, such as a bacterium? Could we, in this case, perfectly isolate the system and obtain a quantum superposition of "alive" and "dead"? At the moment, we do not yet have a definitive answer. The question of understanding to what extent we can observe effects due to quantum superposition in biological or more complex systems remains open and fascinating. In fact, pioneering experiments are already underway to investigate these phenomena on increasingly large objects. In 1999, at the University of Vienna, Nobel laureate Anton Zeilinger and his collaborators demonstrated quantum interference effects for the first time on a fullerene molecule, composed of 60 carbon atoms. Since then, research has continued, led by Markus Arndt, with experiments exploring these effects on increasingly large molecules. The frontier of quantum physics is thus expanding into territories unimaginable until recently.

But what is the connection between this question and the development of quantum computers? We can imagine a quantum computer as a sort of "Schrödinger's cat"—not entirely macroscopic, but not completely microscopic either. Our goal, in fact, is to be able to control as many qubits as possible during the execution of an algorithm to tackle extremely complex calculations. A system like a quantum computer is called mesoscopic because it lies between the microscopic world, represented by individual qubits, and the macroscopic world, where we will never see a cat in a superposition of alive and dead. The real challenge lies in overcoming decoherence, i.e., managing to maintain state superposition in increasingly large systems and for increasingly long times. In other words, to build a working quantum computer, we must win this battle against decoherence. It is a colossal technological challenge, but it is precisely where the future of quantum computing lies.

2.7 Quantum Hardware: Where Are We?

We remember a famous quote by Schrödinger from 1952: "We never experiment with just one electron or atom or (small) molecule. In thought-experiments we sometimes assume that we do; this invariably entails ridiculous consequences [...] we are not experimenting with single particles, any more than we can raise Ichthyosauria in the zoo". If we take these words of one of the greatest and most brilliant founding fathers of quantum mechanics literally, we might be led to think that everything we have discussed so far is just a fascinating mental exercise, perhaps useful for explaining abstract concepts but without any prospect of practical application. In other words, it would be like thinking of "raising Ichthyosauria in the zoo". It is surprising how what Schrödinger considered "ridiculous" became a reality a few decades later. To build a quantum computer, this is precisely what we must do: experiment with and manipulate single particles, such as atoms and electrons, with extraordinary precision. But the technological challenge we face today is even more ambitious. It is not enough to control a single qubit, as in Schrödinger's quote: we must manage entire sets of qubits, maintaining the entangled states of these qubits protected from decoherence. It is a titanic undertaking, but it is precisely on this ground that the future of quantum computing is at stake.

Fortunately, in recent years, technological progress has allowed us to delve into the mysterious and fascinating microscopic world, where distances are measured in millionths of a millimeter. Contrary to what could have been imagined in Schrödinger's time, today we are able to conduct experiments that until a few decades ago seemed unthinkable. It is in the exploration of this extraordinary realm that we must seek the way to build the hardware of a quantum computer. In this world, however, everything is incredibly delicate, and as we venture into the concrete realization of the physical elements of a quantum computer, we must be aware of the challenges and limits imposed by this fragile and complex universe. Here lies the key to the future of quantum technology, but the path requires unprecedented precision and understanding.

The first challenge to address is how to concretely build a qubit. This is not at all simple. A qubit is essentially a quantum system with two distinguishable states, such as an atom that can be in two different states (or orbital levels). These two states, which form the core of the quantum bit, must ideally be isolated from the rest of the universe and possess well-defined characteristics: stability and reproducibility. Fortunately, nature offers us an infinite number of identical atoms (one hydrogen atom is exactly like another), but this is where the complexity begins. Atoms are extremely sensitive to the surrounding environment—so sensitive that, as we will see in Chapter 4, they can be used as excellent sensors! However, this sensitivity also means that small external variations, such as fluctuations in the Earth's magnetic field, can alter the characteristics of one atom compared to another, and consequently, the qubits are no longer perfectly identical. This necessitates the ability to characterize, control, and distinguish systems that are not exactly the same, drastically increasing the engineering complexity. This challenge becomes even more difficult when trying to increase the number of qubits in the computer, a crucial step to making quantum computers truly useful. Take the case of so-called "superconducting qubits". These are nothing more than real artificial atoms, created through printed circuits on a micrometric scale (a thousandth of a millimeter). However, precisely because they are engineered, these qubits are inevitably subject to manufacturing defects and imperfections. Every small imprecision in the manufacturing process can introduce errors, undermining the stability and reliability of the system. And as we aim to increase the number of qubits to make quantum computers more powerful, another great challenge arises: ensuring that all these artificial qubits work in a coordinated and error-free manner, despite the complexity of their construction processes.

Another challenge, apparently simple but in reality far from trivial, is that of correctly initializing a quantum computer, i.e., preparing it for the start of the calculation. Think of when we want to write on a piece of paper: to ensure that our writing is legible, we start with a blank sheet, not one full of scribbles. Similarly, to perform calculations with a computer, we must initialize it in a well-defined state, such as the state representing the number zero, to which we then add other numbers to execute the requested calculation. On an abacus, we move all the beads to a given initial position corresponding to zero; in a classical computer, the memory is initialized with the state 00000..., and in a quantum computer, we must do the same: initialize the qubits to the state 00000..., i.e., in the case of atoms, each must be in a well-defined atomic state. But here, the difficulty arises: initializing these qubits is not at all simple. Take the case of artificial atoms, such as the superconducting qubits we mentioned. Being built inside materials, these are normally at temperatures that cause thermal agitation of atoms and molecules. The higher the temperature, the stronger this agitation, and this means that our qubit—instead of remaining in state 0, the "rest state"—has a certain probability of jumping to the "excited state", i.e., switching from 0 to 1. To avoid this problem, quantum computers must be cooled to extremely low temperatures, close to absolute zero (−273.15°C), so that the probability of atomic excitation is minimized. In addition to the challenges related to the manipulation of atomic systems, there is therefore the complexity of cryogenic technology. Despite progress, this introduces new technical difficulties and challenges yet to be overcome, making the enterprise of quantum computing even more extraordinary.

Suppose we have managed to overcome the enormous challenge of initializing the state of our quantum computer with extreme precision. At this point, we must face another undertaking: implementing a "universal" set of quantum logic gates. This means building a series of operations that, if concatenated in the right way, allow us to execute any algorithm. However, realizing these logic gates is far from simple. Each phase involves a series of experimental obstacles, each with its own practical limitations.

Think, for example, of the maximum power of a laser needed to manipulate atoms, or the duration of a pulse that must not be too long to avoid overheating and damaging the chip. To address these difficulties, optimal control systems are used, similar to those developed for self-driving cars. Just as a car must find the best path to reach a destination, in the case of quantum logic gates, we must "guide" an atom toward the desired state, for example, finding the right laser pulse that takes it exactly where we want. And, just like in autonomous driving, the development of these control systems is increasingly intertwined with that of artificial intelligence, creating a fascinating intersection between advanced technologies that are shaping the future.

The next step is to read the result of the calculation—an operation that, in classical computers, is relatively simple: just access the memory and retrieve the data. In the quantum world, however, reading the final state of the system is not so straightforward and has unique characteristics, as already mentioned. Let's take a practical example of quantum measurement: imagine using qubits represented by ions, i.e., electrically charged atoms, trapped in a region of space. How can we determine if a given ion is in state 0 or state 1? The procedure works as follows: the ion is irradiated with a laser beam of a specific frequency. If the ion is in state 1, the laser excites it, bringing it to a third highly unstable energy level, from which it will quickly decay by emitting a photon, i.e., light. This light can be captured by a special camera. Conversely, if the ion is in state 0, nothing happens. The result? If the ion is in state 1, we will see a bright spot on the screen caused by the emitted light. If the ion is in state 0, our screen will remain dark. This type of reading, while fascinating, reflects the delicate and complex nature of measurement in the quantum world, where even observing a single photon can make the difference between the success or failure of a quantum calculation.

Finally, a last and crucial requirement is that the quantum system maintains its nature for the entire time necessary for the calculation. Decoherence, i.e., the loss of quantum properties caused by the inevitable interaction with the external environment—must not compromise the computer before the process is completed. In other words, the decoherence time, i.e., the moment after which the quantum computer loses its quantum mechanical advantages—must be longer than the time required to execute the calculation. If this does not happen, external interference destroys the delicate superpositions and entanglement, canceling the quantum potential of the system. Without superposition and without entanglement, a quantum computer is no more efficient than a classical computer, and all efforts to develop it would be in vain. Its success, therefore, depends on the ability to protect it from these interferences for the entire time required by the calculation—a challenge that is both fascinating and fundamental.

This long list of requirements to be met, or rather challenges to be overcome, may seem almost discouraging, and undoubtedly the reader has grasped the enormous complexity inherent in building quantum computers. Yet, the progress made in recent years has been extraordinary, beyond all expectations.

In the following, we want to take a look at some of the most promising architectures being developed in this field. It is likely that within a few years, the quantum computers available to us will have made further significant strides, and it is equally reasonable to think that the foundations on which these systems are based will remain solid for a long time. Evolution is rapid, but the foundations are destined to last.

2.7.1 *Trapped ions*

In a late-century summer, in the majestic mountains of Colorado, an international conference on atomic physics was held. On that occasion, the scientific community learned of two significant discoveries. Peter Shor had just presented his revolutionary quantum algorithm for integer factorization, a result that threatened to make existing cryptographic systems obsolete. Around the same time, Artur Ekert had developed a protocol for transmitting information securely, leveraging the unique properties of quantum mechanics. These discoveries opened the door to disruptive scenarios in the world of technology and information security.

Peter Zoller and his young collaborator, Ignacio Cirac, present at the conference, put forward an idea that would trigger a global race to build the quantum computer. Their proposal? Use ions, i.e., electrically charged atoms, confining them in special "traps" within a region of space where a vacuum is created and cooling them to extremely low temperatures. Two energy levels of an ion, corresponding to two different orbits of an electron, can be used to represent the two states 0 and 1 of a qubit. But the real key is that, being charged, the electrostatic interaction acts between them. This interaction generates entanglement, which is the necessary condition for the operation of a quantum computer. However, ions, having the same charge, tend to repel each other and move away. How to solve this problem and keep them trapped in a precise region where they can be manipulated?

The answer lies in the use of electric fields generated by electrodes, with blades and tips (as illustrated in Figure 2.2), which constitute the trap and allow the ions to be kept under control, paving the way for their manipulation and quantum computing.

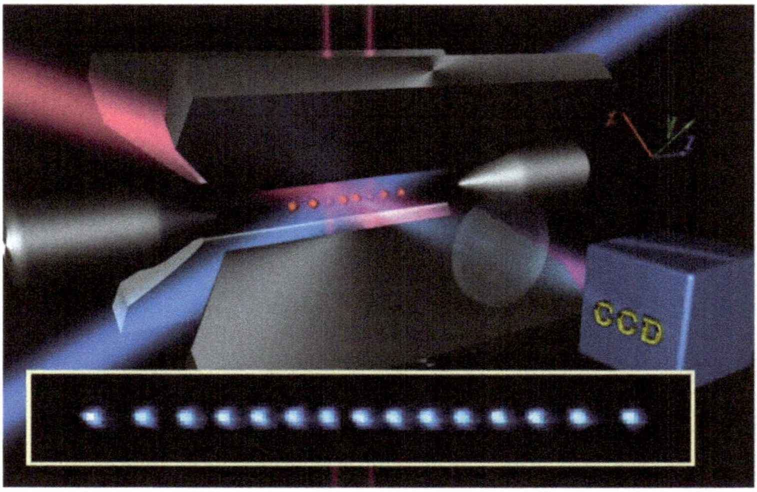

Fig. 2.2. Trapped ions (magnified below) by electric fields (plates and tips in the figure). The manipulation and measurement (the latter through a special camera, called CCD) of the ions' state are done by sending laser light (courtesy of Rainer Blatt, University of Innsbruck).

Imagine a basin full of balls: normally, these would settle at the bottom, grouping together. But ions behave differently: being charged, they repel each other. Instead of clumping, they find a balance, arranging themselves at a certain distance from each other and forming an ordered structure known as an "ion crystal". In this scheme, each ion represents a qubit. Logic gates are realized with laser pulses: for each, an appropriate sequence is defined. Thanks to the electrical interaction between them, the qubits can interact, enabling operations such as two-qubit logic gates, for example, the CNOT gate. In conclusion, by concatenating sequences of laser pulses, any quantum algorithm can be executed.

Since ion trapping and cooling were already well-established technologies, within a year, in 1995, researchers from David Wineland's group (Nobel Prize in Physics in 2012) at the National Institute of Standards and Technology (NIST) in Boulder, Colorado, succeeded in realizing the CNOT logic gate using beryllium ions. A fundamental step had been taken.

Since then, progress has been extraordinary. Many of the complex problems mentioned earlier have been, if not completely solved, at least sufficiently overcome to allow the construction of quantum computer prototypes. Today, in addition to the numerous research laboratories working tirelessly to perfect this technology, there are companies and startups born in an academic context, and even research centers, that sell trapped-ion quantum computers or offer access to such machines. Although these computers are capable of performing non-trivial calculations, they remain limited by the relatively small number of ions—and thus qubits—that can be trapped and manipulated with the required precision. However, these steps mark the beginning of a new era in computing, where the dream of quantum computing is becoming increasingly concrete.

Currently, quantum computers can manage at most a few dozen qubits and are capable of executing only a few dozen logical operations in sequence, such as CNOT gates, before decoherence intervenes and compromises the potential of the calculation. To get an idea of how far we still are from the performance of classical computers, just think of the classical supercomputer Frontier, at Oak Ridge Laboratory, which in May 2022 crossed the exaflop barrier. This means that it is capable of executing one billion billion (i.e., 1 followed by 18 zeros) operations per second.

The road ahead is still long, but the progress made in such a short time gives a glimpse of a promising future. The current limitations of quantum computers do not discourage researchers, who continue to explore increasingly ingenious solutions to overcome the challenges of decoherence. One of the most promising ideas is that of so-called "segmented traps". This approach involves integrating multiple traps on the same chip, each containing a few dozen ions, with the possibility of making them "travel" between the different traps. Only a few years ago, the idea of moving single qubits (ions) quickly and without altering their state would have seemed like science fiction. Today, however, it is a concrete reality under discussion, demonstrating how rapidly the field of quantum computing is evolving. These advances offer an optimistic glimpse into a future where current limitations could be overcome.

2.7.2 *Superconductors*

When an electric current flows through a wire, it encounters resistance: the wire heats up and can even become incandescent. It is as if the material opposes the passage of electrons, trying to block their flow. To maintain the current, energy must be continuously supplied, and this obviously comes at a cost. However, there are special materials called superconductors that, below a certain temperature, allow electric current to flow without encountering any resistance. In these materials, currents can be maintained for very long periods and can reach extremely high intensities. This extraordinary phenomenon is already being exploited in many technologies, thanks to the fact that electric currents generate magnetic fields. Such magnetic fields are used, for example, in Japan and China, in magnetic levitation trains, where superconductors cooled to very low temperatures (mounted on board special cryostats) lift the train. The same technology is at the basis of high-precision Nuclear Magnetic Resonance (NMR) machines used for clinical diagnostics.

Unlike normal electric wires in our homes, in superconductors, it is not individual electrons that flow through the material to form the electric current. In these materials, when cooled below a certain critical temperature, something extraordinary happens: electrons pair up, forming so-called "Cooper pairs", named after Leon Cooper, one of the scientists who contributed to the discovery of superconductivity and won the Nobel Prize in Physics in 1972. These Cooper pairs are able to flow through the material without encountering resistance, making superconductivity possible.

Technology based on superconductors differs from that of trapped ions in that, instead of directly using real atomic systems, it employs circuits that behave like real "artificial atoms". These circuits allow the creation of quantum state superpositions and entanglement, essential elements for quantum computing. In this approach, qubits are realized by leveraging Cooper pairs: circuits made of superconducting materials are printed, where some sections act as small boxes (called "Cooper Pair Boxes"), inside which Cooper pairs can enter in a controlled manner. The qubit is then defined by the presence or absence of a Cooper pair inside such boxes. Through appropriate circuits and electric pulses, it is possible to manage and manipulate these qubits, thus making superconducting circuits an extremely promising platform for quantum computers.

Currently, some of the largest multinational companies are exploring this path to develop the quantum computer of the future. An extraordinary example is Google's Sycamore chip, depicted in Figure 2.3, the result of work by their laboratories led by a pioneer in the field, John Martinis—he was awarded the 2025 Nobel Prize in Physics, together with John Clarke and Michel Devoret, for experiments carried out in the 1980s that demonstrated that quantum effects are not limited to microscopic particles, paving the way for superconducting qubits. The Sycamore chip, composed of 53 qubits (originally 54, but one was defective), led Google to announce in October 2019 that it had achieved the much-discussed "quantum supremacy", i.e., it had performed a calculation practically impossible for any classical supercomputer. Meanwhile, based on the same technology, IBM has developed an impressive infrastructure for the future development of quantum algorithms and programming languages. Moreover, with great educational impact, IBM allows universities to remotely implement algorithms on their quantum computers, accessible via cloud through a web interface.

Fig. 2.3. Google's Sycamore chip (top) and the cooling apparatus of an IBM superconducting quantum computer (bottom). Figures are taken from `https://www.nature.com/articles/s41586-019-1666-5` (top) and from `https://www.ibm.com/quantum` (bottom).

Looking to the future, the superconducting architecture still has to address numerous challenges. The first, and perhaps the most significant, is the need to maintain extremely low temperatures for quantum computers to function. This makes it unlikely that we will have such machines directly on our desks; instead, they are likely to be hosted in specialized computing centers, accessible via cloud. The constraint of low temperatures is not in itself an insurmountable obstacle; the real problem is that it involves high costs and makes the engineering of circuits much more complex, also complicating scalability, i.e., the ability to increase the number of qubits. However, despite these difficulties, the expectations of multinational companies working on this technology are extremely ambitious. These companies plan to reach thousands of qubits in the next few years, suggesting that the quantum revolution could influence our lives much sooner than the more cautious predictions had imagined.

2.7.3 *Rydberg atoms*

We conclude this overview of the various architectures competing to build quantum computers by talking about a new contender, an outsider that has only recently entered the race but is advancing rapidly. In this case, qubits are formed by neutral atoms, lacking electric charge. To manipulate these atoms, they are brought into extreme excitation states, creating what are known as "Rydberg atoms", named after the Swedish physicist Johannes Rydberg, who lived at the turn of the nineteenth and twentieth centuries. Rydberg studied what happens to atoms when one or more electrons are pushed into orbits very far from the nucleus. To make an analogy, it is as if the Earth were moved to an orbit thousands of times farther from the Sun, even farther than Pluto.

To give an idea, consider that the size of a typical atom is on the order of a nanometer (a millionth of a millimeter), while the size of a Rydberg atom is on the order of a micron (a thousandth of a millimeter), i.e., the atoms effectively become a thousand times larger. Rydberg atoms retain all the quantum properties we want to exploit for performing quantum calculations, but they have much larger dimensions, and therefore are much more easily manipulable.

For example, they can be trapped and moved with extreme precision using tools called "optical tweezers", which use laser beams to guide single Rydberg atoms into precise positions, with a resolution on the order of a micron. This technique is so refined that some laboratories have succeeded in forming recognizable figures, as if the single Rydberg atoms were pixels. In Figure 2.4, we show some of these fascinating results obtained in Antoine Browaeys' laboratories in Paris.

Recent progress in this technology, which could revolutionize the development of quantum computers, will be explored in Chapter 5. Furthermore, systems based on Rydberg atoms are also at the forefront in another crucial application of quantum physics: quantum simulations.

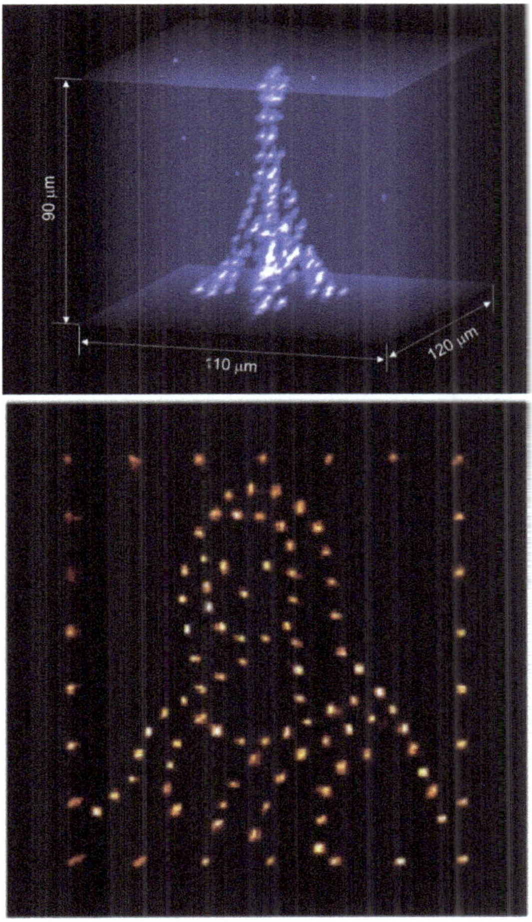

Fig. 2.4. Rydberg atoms in artistic configurations obtained by positioning the atoms (the dots in the figures) using optical tweezers. Figures are taken from `https://www.usinenouvelle.com/article/calcul-quantique-pasqal-s-impose-dans-la-cour-des-grands-avec-ses-atomes-froids.N1809802` (top) and from `https://journals.aps.org/pra/abstract/10.1103/PhysRevA.102.063107` (bottom).

2.8 Quantum Simulations

The optimization of production processes has now reached incredibly complex levels. Think, for example, of optimizing the aerodynamic properties of spacecraft, airplanes, racing boats, or racing cars. Many cutting-edge sectors are increasingly dependent on our ability to improve incredibly complex systems and predict their behavior even in extreme conditions: from racing-car components designed to endure extreme temperatures, to buildings engineered to withstand earthquakes.

But it is not just about advanced technology. Many of our daily decisions are also based on the ability to make accurate predictions. Think of the weather: in the short term, it influences our weekend trips and civil protection alerts; in the medium term, it determines harvests and investment strategies in the agri-food sector; in the long term, it shapes our responses to climate change. In equally crucial areas, the ability to design increasingly efficient antennas, and thus understand how electromagnetic fields propagate in the atmosphere or in materials, influences the development of new mobile phones, computer chips, and communication infrastructures. And, further, understanding how atoms combine into molecules, macromolecules, and finally proteins or enzymes is fundamental to accelerating the creation of new materials, drugs, and vaccines. Our ability to optimize and predict thus becomes the key to addressing the technological and scientific challenges of our time.

In general, the systems just described, like most phenomena of scientific, social, and economic interest (think of migration flows or stock markets), are inherently complex. Their dynamics, i.e., their evolution over time—is often extremely difficult to predict. The calculations necessary to anticipate the behavior of these systems cannot be done simply with pen and paper, but require the use of powerful computers. However, these computers, to function, depend on specific algorithms designed to solve the problem. In many cases, however, the available algorithms are so complex that they require incredibly long computation times, which can even exceed the capabilities of the world's most advanced supercomputers. This explains, for example, why we are unable to predict the weather accurately beyond a few days. The complexity of the systems and the limits of our current technology make these predictions an undertaking still beyond our reach.

In some situations, there is a completely different way to overcome these difficulties—an approach that can lead to much faster and more efficient results. This involves simulating the system of interest in a controlled environment and observing how it behaves. In practice, an accurate model of the system we want to study—our simulator—is created and used to test and improve existing technologies. An iconic example of this approach is the wind tunnel: to develop new airplanes, boats, or cars, and optimize their performance, we analyze how air or water interacts with the vehicles under various conditions. In theory, the motion of an object moving through a fluid, such as water or air, can be described precisely by well-known mathematical equations: the equations of fluid dynamics. However, solving these equations for a realistic system is a titanic undertaking that requires enormous computing power. The alternative, then, is to build a wind tunnel, where a scale model of a new airplane, for example, is tested under simulated atmospheric turbulence conditions. In this way, the problem is not addressed through a generic computer, no matter how powerful it may be, but through a physical simulator—in our case, the wind tunnel. This tool is capable of solving a specific problem with great efficiency, providing quick and detailed results that would otherwise be difficult to obtain. Thanks to these simulations, we can extract valuable information without having to resort to extremely complex numerical calculations, significantly accelerating technological progress. The final result is to make vehicles faster or more energy-efficient, improving every aspect of their performance.

The idea of developing a simulator, rather than a universal computer capable of solving a wide range of problems, also finds powerful application in the quantum world. As we have said, building a real quantum computer is extremely complex, and the challenges involved in simulating complex systems will remain formidable for many years to come. For this reason, a potentially more effective path is to create a simpler machine: a quantum simulator, capable of replicating the specific behavior of a particular system, thus allowing us to study it in depth. An extraordinary example of this idea was realized at the end of 2020, using simulators based on Rydberg atoms. These tools were able to reproduce the behavior of materials composed of over 200 atoms—an undertaking at the limits of impossibility for any classical supercomputer.

One of the most promising areas for the application of quantum simulators, and perhaps the most natural, is the study of materials and their extraordinary properties. We have already mentioned superconducting materials as key elements for realizing qubits, but what we have not emphasized is that we still understand little about the mechanisms underlying superconductivity, especially that near room temperature. This gap limits our ability to optimize such materials to improve their performance. Here, the potential of quantum simulators comes into play: in the near future, they could become fundamental tools for improving our understanding of superconductivity and accelerating the development of advanced materials, with applications that could strongly impact entire sectors. For example, these simulators could tackle complex problems in chemistry, where traditional simulations with classical computers of very complex molecules become almost impossible to handle. The ability to replicate complex chemical reactions through quantum simulations could open the way to surprising discoveries in chemistry and medicine.

In recent years, a new and fascinating line of research has emerged: building quantum simulators to explore the properties of the fundamental constituents of the universe, elementary particles. So far, one of the most extraordinary tools for studying these particles has been CERN's Hadron Collider, a technological giant that made possible the discovery of the Higgs boson in 2012, the famous "God particle". This colossal undertaking required decades of work, the contribution of thousands of physicists and engineers, and the construction of a ring to accelerate particles along 27 kilometers, together with state-of-the-art technologies to make it work and analyze its data. Inside this accelerator, particle beams collide at vertiginous speeds, generating the reactions that physicists want to study. Today, however, some scientists are aiming for something even more surprising: a quantum simulator to study collisions between particles. It is an extremely ambitious project that will require years of development to reach the maturity necessary to produce useful results for research. But the idea of being able to replace CERN's colossal experiment with an experiment conducted on a simple optical bench in a normal room is truly fascinating.

Another equally fascinating problem, for the study of which quantum simulators are proving useful, is the exploration of the mysterious boundary between the classical and quantum worlds. In the first part of this book, we introduced concepts such as superposition and entanglement, emphasizing that these properties cannot exist in the classical world. But where exactly is the limit between these two realms, classical and quantum? An atom clearly operates in the quantum world, while we, made of billions and billions of atoms, live in the classical world. Between these two extremes, however, some kind of transition must occur, whose nature remains shrouded in mystery: is it sudden or gradual? For example, where do ten atoms fit? And a cell? It is likely that a set of ten atoms belongs to the quantum world, while a cell seems to belong to the classical world.

This enigma confronts us with a deeply intriguing question. We mentioned how Anton Zeilinger's group demonstrated years ago that it is possible to observe quantum superposition effects in a fullerene molecule composed of sixty carbon atoms. Today, some researchers are trying to push the boundaries even further, attempting to create conditions in the laboratory to put a bacterium in a superposition of two distinct states.

Finally, in the world of quantum research, a real "race" is underway to create the largest possible quantum superposition in the laboratory—a sort of giant "Schrödinger's cat" composed of the largest number of qubits. In 2020, using a quantum simulator based on Rydberg atoms, an impressive milestone was reached: the creation of an entangled state of twenty qubits. This result represents a milestone, but we are only at the beginning. The race continues and promises to hold surprises, because the ability to create quantum states with an increasing number of qubits is fundamental to performing increasingly complex calculations. The ultimate goal? A clear and irrefutable demonstration of the superiority of quantum computers over classical ones: quantum supremacy.

2.8.1 *Quantum supremacy?*

In the autumn of 2019, an unexpected leak of information from a NASA data center erupted in the academic world and, shortly after, reached the general public: Google was ready to announce that it had achieved the long-awaited "quantum supremacy". For a decade, Google had worked tirelessly toward this result, launching an impressive hardware and software development program that, in a few years, had placed them at the forefront of the race to develop the quantum computer of the future. In mid-2019, Sycamore was presented, the 53-qubit chip, a promising step forward for quantum computing. However, having a cutting-edge chip was not enough: it was also necessary to demonstrate the ability to perform a calculation that no classical supercomputer in the world could replicate—an undertaking made complex by the limited number of qubits and the problem of decoherence.

To overcome this challenge, Google's team selected a specific algorithm—one with no immediate practical applications, yet complex enough to test the chip's capabilities. The algorithm involved applying a random sequence of logic gates, generating an extremely complex quantum state. To replicate the same calculation on a classical computer, it would be necessary to define 2^{53} numbers (these numbers are required to specify a generic state of a quantum system with 53 qubits). This amount of information is equivalent to writing over ten million billion numbers, which would occupy several dozen petabytes of memory. Furthermore, verifying the correctness of this quantum state would require performing an astronomically large number of operations, further increasing the complexity of the problem. Google's team estimated that a classical supercomputer would take approximately 10,000 years to complete the same calculation that the Sycamore quantum chip finished in just a few minutes. With this extraordinary result, quantum supremacy seemed to have finally been demonstrated, marking a historic turning point in the field of information technology.

Google's announcement caused quite a stir and will likely be remembered as a turning point in the development of quantum computers. However, shortly thereafter, one of Google's main competitors, IBM, claimed that their classical supercomputers could replicate the same calculation within a few days, thereby tempering the initial enthusiasm. It should be noted, though, that this claim was never supported by a concrete demonstration or verifiable calculations, and thus remained merely a statement of feasibility.

More interestingly, a scientific paper published in 2020 showed that, through methods of approximation and information compression (based on so-called tensor networks, developed by physicists to simulate quantum systems), it was possible to accurately reproduce Google's results using an ordinary laptop computer within just a few hours. This study opened up new perspectives, demonstrating that "quantum supremacy" is not as clear-cut as it might seem—at least for now.

Regardless of the debates surrounding the actual achievement of quantum supremacy, one thing is certain: the development of quantum hardware is advancing rapidly. Although the term "quantum supremacy"—increasingly replaced by "quantum advantage" to avoid uncomfortable historical connotations—may not yet have been fully demonstrated, it is hard to deny that we are witnessing the dawn of a new technological era. If the so-called Neven's Law, formulated in 2018 by Hartmut Neven, director of Google's quantum computing program, proves to be correct, we can expect an extraordinary acceleration in the development of quantum computers in the near future. This law predicts that the computational power of quantum computers will grow as a double exponential (i.e., an exponential of an exponential—far faster than the simple exponential growth illustrated by the legend of the pharaoh and the chessboard). If this trend continues, in the coming years quantum computation and simulation will not merely serve as advanced research tools but will radically transform our world. However, as often happens in technological development, the future remains unpredictable: Neven's Law is based on empirical observations, and it is difficult to foresee how long it will remain valid.

In 2020, a team led by Jian-Wei Pan at the University of Science and Technology in Hefei, China, announced a new experiment that claimed to have achieved the long-sought quantum advantage—this time by solving a specific problem known as Boson Sampling. Unlike other experiments, this one employed a different type of quantum hardware: photons manipulated within an optical circuit.

The use of photons is also becoming a standard technique in another crucial area of quantum technology—secure communications—which promise to be immune to any form of hacking attempt. We will explore this topic in the next chapter.

Chapter 3

Quantum Communications: The New Frontier of Security

> It may be roundly asserted that
> human ingenuity cannot
> concoct a cipher which human
> ingenuity cannot resolve.
>
> Edgar Allan Poe, *The Gold-Bug*

3.1 At the Roots of Security: Classical Cryptography

In the 5th–4th centuries BCE, Greece experienced a golden age, characterized by a great flourishing of art, literature, and philosophy. Figures such as Plato, Aristotle, Euclid, and Hippocrates dominated the intellectual landscape of the time. The great tragedies of Aeschylus, Sophocles, and Euripides were performed in the theaters of Athens. It was in this cultural and political context that Herodotus lived, one of the greatest historians of antiquity, known for recounting the history of the Persian Wars against Greece. Even in those times, wars were fought not only with weapons but also with other more or less sophisticated means. As Herodotus narrates, messengers were entrusted with delivering communications, and these messengers often had to cross hostile territories, risking interception and capture. It was necessary to protect the content of the messages.

A technique for concealing a message is described by Herodotus in Book V, 35 of his Histories (written around 440–429 BCE):

> ...just at that moment, a messenger arrived from Susa, sent by Aristagoras, with signs tattooed on his head urging Aristagoras to rebel against the king. For Istiaeus, wanting to send a message to Aristagoras ordering him to rise up, had no sufficiently secure way to warn him, as all roads were under control; so he shaved the head of his most trusted servant,

> tattooed signs on it, waited for the hair to grow back, and as soon as it had grown, sent him to Miletus with the sole mission, once he arrived there, to ask Aristagoras to shave his head and take a look at it.

Herodotus also recounts (Book VII, 239) that Demaratus, a Greek exile at the Persian court, warned Sparta of an imminent invasion by Xerxes, the Persian king, in the following way: he removed the wax from a tablet, carved his message on the underlying wood, covered it again with wax to hide it, and finally sent the tablet which, appearing unused, was able to reach its destination.

These early attempts to protect the secrecy of a message are based on the idea of hiding the message itself. This technique is called steganography, from the Greek steganos (covered or hidden) and graphein (to write). A more recent (and less laborious) method of steganography is described by the Italian philosopher and scientist Giambattista Della Porta in his work De Furtivis Literarum Notis (1563): it consists of writing the message on the shell of a hard-boiled egg using as "ink" a mixture of one liter of vinegar and 60 grams of alum. Being porous, the shell allows the ink to penetrate without leaving any external trace. The recipient of the hard-boiled egg can peel it and read the message

Steganography played an important role in ancient times and has continued to evolve over the centuries, improving with technological progress. A more sophisticated technique, used in recent times, is that of the microdot. It involves reducing the size of a text through photographic processes to less than a millimeter. The message can thus be hidden under a dot or a comma in a normal letter. The German intelligence agency Abwehr regularly used microdots during World War II to send secret messages. This method proved very effective and was only discovered through espionage activities.

In today's increasingly interconnected society, the need to communicate securely is growing and now permeates our daily lives. All of us want to communicate securely online and prevent our messages from being read by third parties. This is especially true in the case of monetary transactions, when we apply a digital signature, etc. To protect our privacy and safeguard personal data, especially sensitive ones, from prying eyes, the universally used method today is cryptography—the art and science of secret communication. Cryptography is a process that alters information, for example that contained in a message, to render it meaningless to anyone who does not possess the key to decrypt it. It is not, therefore, a way to prevent cyberattacks, but a technique to make data unusable for those who intercept it improperly.

Cryptography (from the Greek kryptos, hidden) is thus unlike steganography, based on the idea of altering the message so that its meaning is not understandable. It has a long and fascinating history, dating back thousands of years to the ancient civilizations of Egypt and Mesopotamia. The famous Code of Hammurabi, for example, a set of laws created in Babylon around 1754 BCE, made reference to the use of secret messages to transmit information. Historians believe that the Babylonians used a simple substitution cipher, where each letter of the alphabet was replaced by another letter or symbol, to encode their messages.

The Atbash cipher, used in the Book of Jeremiah, reverses the order of the letters in the Hebrew alphabet: the first letter of the alphabet (aleph) is swapped with the last (tav), the second (beth) with the penultimate (shin), and so on; the word "Atbash" itself seems to be an acronym composed of these four letters "Aleph-Tav-Beth-Shin". Using this cipher, in the Book of Jeremiah, Babylon, the capital of the Babylonian Empire, becomes Sheshach. In fact, the letters of "Sheshach" can be rearranged to form the letters of "Babel" or "Babylon" (gate of God). To avoid misunderstandings, however, it is important to emphasize that the Book of Jeremiah, one of the books of the Hebrew Bible and the Christian Old Testament, is primarily a collection of prophetic oracles, poems, and historical narratives attributed to the prophet Jeremiah. The main purpose of the text was to convey its prophetic message to the public, rather than hiding its meaning through cryptography. The use of the Atbash cipher represents a relatively rare example of cryptography used in ancient Hebrew literature.

But the origins of cryptography date back even further in time. A traveler who arrives in Beni Hasan, on the eastern bank of the Nile in Middle Egypt, can visit an impressive complex of tombs. Among these stands out the magnificent Tomb 3, that of Khnumhotep II, a nobleman who lived during the reign of Pharaoh Amenemhat II, around the 20th century BCE. The walls of the tomb are decorated with paintings depicting scenes from the life of Khnumhotep II and, in general, of the society of the time. These paintings are characterized by remarkable realism and attention to detail, offering a vivid description of the life and culture of ancient Egypt and also providing an idea of the social hierarchy prevalent at the time. At first glance, the hieroglyphs in these paintings seem to be simple decorations. However, upon closer inspection, it is noted that some of them are written backwards or upside down, and some of the symbols are replaced by others with different but similar meanings. We are thus faced with an example of cryptography. The reason why the Egyptians felt the need to encrypt these inscriptions was probably only to hide information about their religious ceremonies from the common people. Or simply as an aesthetic choice or to impress visitors. Other examples of cryptography have been found in tombs and inscriptions from this period, suggesting that it was an important tool for protecting secrets and knowledge in ancient Egyptian society. The cryptography used by Khnumhotep II and the ancient Egyptians is relatively simple by modern standards. However, it demonstrates the early human desire to protect information from unauthorized access.

Cryptography has naturally also had extensive military use. For example, during the Peloponnesian War, which pitted the great powers of Sparta and Athens against each other, the Spartans sent encrypted messages using the scytale (or skytale), a rod around which a parchment was wrapped in a spiral, which was then used to write a text along the axis of the rod, line after line (see Figure 3.1). The parchment, when unwound, presented a sequence of letters apparently meaningless, and was then sent to the recipient, who possessed a rod identical to that used to encrypt it, around which it was wrapped to make the original message reappear.

Fig. 3.1. Example of a scytale. The inscription visible on the wrapped strip is in Italian ("LA SCITALE/E' UN ANTICO"), which may be translated into English as "THE SCYTALE/IS AN OLD...".

As Plutarch recounts in the Life of Lysander, after the Peloponnesian War, around 400 BCE, the Persian satrap Pharnabazus, an ally of Sparta, complained to the ephors (the most important magistrates of Sparta) about the looting perpetrated by the Spartan navarch (admiral) Lysander and his lieutenant Thorax:

> The ephors, angered at first, killed his friend and comrade-in-arms Thorax, because they found silver in his house, and to Lysander himself they sent the scytale, ordering him to return. The scytale is made as follows: when the ephors send an admiral or general to the armies, they polish two round sticks of equal thickness and length, and their cuts match perfectly with each other: one they keep, and the other they give to the

> citizen they send; and these sticks are called scytale. And when they want to communicate some important matter to be kept secret, they take a long and narrow strip of bark like a thong and wrap it around their scytale in a spiral without leaving any space of wood visible, but covering the entire surface: and having done this, they write on that wrapped strip or paper what they want; and having removed the paper, with the writing on it, they send it to their captain without the stick: who, upon receiving it, cannot read the scattered letters without any connection: but taking his scytale, he stretches it out and wraps the paper around it in the same order as before, and thus finds the continuous writing around it, and the paper is also called scytale by the same name as the stick, just as the measured body is called by the name of the measure.

Despite being successfully used by the Spartans, the scytale certainly did not guarantee total security: once its operation was discovered, enemies could build a stick of the same shape and decrypt the message.

A more sophisticated cryptographic system was devised by Julius Caesar. During the Gallic Wars, communication with the troops was made difficult by the impossibility of maintaining full control over the communication routes in foreign territory. Messages could be easily intercepted, and the enemy could learn of strategic military information. To solve this problem, Caesar devised a cipher that bears his name: the "Caesar cipher". It consists of shifting the letters of the alphabet by a fixed number of steps (which constitutes the key of the code). For example, using a key of 3, A becomes D, B becomes E, C becomes F, and so on; when reaching the end of the alphabet, we have to count 3 steps starting from the beginning, for example Z becomes C. With this encoding, the message "OMNIA GALLIA EST DIVISA IN PARTES TRES" is encrypted as "RPQLD JDOOLD HVW GLYLVD LQ SDUWHV WUHV". To decrypt it, it is sufficient to possess the key, in this case 3, and count 3 steps back for each letter: R becomes O, P becomes M, Q becomes N, and so on.

To make the cipher more complex and difficult to decrypt, variants have been invented. An interesting variant of the Caesar cipher replaces the letters with one of their possible permutations, i.e., the order of the letters in the alphabet is changed, as in anagramming a word. If we have two letters, the possible permutations are two: AB and BA; with three letters, they become six: ABC, ACB, BAC, BCA, CAB, CBA; with the 26 letters of the English alphabet, the number of permutations becomes enormous.

Nevertheless, even these variants of the Caesar cipher are easy to decrypt today, because the letters of an alphabet appear with different frequencies in a text. For example, vowels appear more often than consonants. Therefore, to decrypt the code, a simple statistical analysis of the encrypted text is sufficient. However, despite its simplicity and vulnerability, the Caesar cipher remains a fascinating attempt to protect sensitive information. It represents a crucial point in the evolution of cryptography and continues to be a valuable educational tool for introducing students to the complex world of cybersecurity. For example, it can be easily implemented on a computer as a useful exercise for a student.

The idea of encrypting a message through permutations of the alphabet is found in the beautiful cipher disk of Leon Battista Alberti, described in his treatise *De cifris* (1467–1469), one of the first texts on cryptography. Alberti was a key figure in the cultural landscape of the Italian Renaissance. His life was characterized by intellectual curiosity, the pursuit of knowledge, and commitment to creating works of great artistic and cultural value. In addition to writing a treatise on cryptography, he was also an architect, writer, mathematician, and musician. The idea of his cipher disk (see Figure 3.2) is roughly as follows: the letters of the alphabet used for the plaintext are written in uppercase on an outer disk. On a rotating inner disk, the letters of the alphabet are written in lowercase but in a disordered way. These are used for the ciphertext. The inner disk also has a special symbol ("et" in the figure). A key is established, for example the letter I. The inner disk is rotated so that the special symbol aligns with the letter I on the outer disk. This creates a correspondence between the uppercase letters used for the plaintext and the lowercase inner letters used for the ciphertext. The key, that is the uppercase letter, is then frequently changed during the encryption of the message, rotating the inner disk each time to position the new key in correspondence with the special character. Essentially, Alberti no longer uses a single encrypted alphabet to encode a message but a polyalphabetic system, with the alphabet changing each time the uppercase letter is changed, thus making the code much more difficult to break. The recipient, in possession of the same disk, can easily read the message. It is sufficient that, each time a capital letter is received, they rotate their inner disk so that this letter becomes the new key, to then continue the decryption.

Alberti's invention was fundamental for the development of more advanced cryptographic techniques in the following centuries. It is worth recalling the sophisticated "Jefferson cipher", invented by Thomas Jefferson (1743–1826), third president of the United States of America, one of the main architects of the American Revolution, author of the United States

Fig. 3.2. Example of Leon Battista Alberti's cipher disk. The correspondence between the letters of the plaintext and those of the encrypted message is obtained by positioning the special character "et". Here "et" corresponds to the letter I, which is the key of the code. Knowing that, we know that the letter L is encoded as m, M as z, N as d, and so on.

Declaration of Independence adopted on July 4, 1776 (incidentally, a curious fact about his personality: in this declaration, Jefferson speaks very eloquently in favor of the natural rights of man and, in his writings, states that slavery is a political and moral evil; however, during his life, he purchased slaves and at his death had over six hundred on his Monticello plantation in Virginia).

Jefferson was fascinated by cryptography throughout his life and, even during his presidency (1801–1809), personally worked on developing secret codes for diplomatic communications and military purposes.

Jefferson's wheel (or cylinder) consists of wooden disks threaded onto a metal rod. The letters of the alphabet are engraved along the edge of each disk, using a different permutation for each. The disks are then rotated until the plaintext message appears on a line (for example, in Figure 3.3 "COOL JEFFERSON WHEEL CIPHER"). The sequence of letters on another line is then sent (for example, in the figure, two lines above the plaintext we have "IECHTFA..."). This sequence of letters constitutes the ciphertext message that can be decrypted by the recipient in possession of a wheel with the same sequence of disks (i.e., the same permutations of the alphabet), aligning them so that the ciphertext message appears on one line and, at that point, reading the plaintext message on another line.

Fig. 3.3. Example of a Jefferson cylinder, used as a polyalphabetic cipher. Each disk contains a permutation of the letters of the alphabet. On one line, the plaintext message appears ("COOL JEFFERSON WHEEL CIPHER"); any other line can be used as the encrypted message, easily decipherable by anyone who has a Jefferson cylinder with the same sequence of disks (permutations of the alphabet). Figure taken from `https://www.monticello.org/research-education/thomas-jefferson-encyclopedia/wheel-cipher/`

Jefferson's cylinder was the basis of the famous M-94 cipher machine, then considered unbreakable, which remained in service until 1942. We recall here that a cipher machine automatically transforms the letters of the original plaintext into the letters of the ciphertext and vice versa.

3.2 Cryptanalysis

So far, we have talked about the development of cryptography over the centuries and the progressive improvement of techniques for encrypting a message, i.e., hiding its meaning. This process, in turn, has stimulated the study (known as cryptanalysis) of techniques for decrypting a message and reading its meaning without possessing the key. The history of cryptography is, in fact, the long history of a battle between, on the one hand, finding the best way to hide the meaning of a message and, on the other, searching for new and more sophisticated ways to discover its meaning. This battle is still ongoing. Cryptanalysis evolves hand in hand with cryptography, which is therefore forced to improve, and vice versa.

During the period of the medieval Arab civilization, which extends roughly from the 7th to the 13th century, there were important developments in cryptanalysis and cryptography. In this period of great flourishing of Arab civilization, Muslim scholars made remarkable progress in various scientific disciplines, including mathematics, algebra, and number theory, and this contributed to the improvement of encryption and decryption techniques. Around the year 1000, in Saudi Arabia, the first frequency analysis was born, capable of breaking codes based on monoalphabetic substitution. Until the High Middle Ages, in fact, cryptographic systems based on shifting or permutation continued to be used. They had evolved so much that it seemed impossible to decrypt them. However, they had a weak point: the letters in a message appear with different frequencies (for example, in a text in English, A, T, and E appear more often than other letters). By analyzing the frequency of letters in an encrypted message (using monoalphabetic substitution) and comparing them with those expected for the letters of a plaintext message, it was possible to associate each encrypted letter with the plaintext letter that had the same frequency, thus arriving at the decryption of the text. This cryptanalytic activity began with the in-depth study of the Quran, for which scholars sought to determine, along with many other characteristics, the frequency of use of the letters.

As early as the 9th century, the great Arab philosopher and mathematician Al-Kindi published "Epistle on the Decryption of Encrypted Messages". This work gave rise to the first true form of cryptanalysis: he proposed examining a sufficiently long plaintext and ordering the letters according to the frequency with which they appear. By doing the same for the encrypted message, the correspondence between encrypted letters and letters before encryption is found, and this allows the message to be decrypted. The vast Arab knowledge in the field of cryptology is documented in the corresponding section of a voluminous encyclopedia completed in 1412, in 14 volumes: "The Dawn of the nearsighted in the Art of Composition" by al-Qalqashandī, an Egyptian historian and mathematician who was a scribe attached to the Mamluk chancellery in Cairo. It is worth emphasizing that these works testify to the connection between scientific culture and literary culture present in Arab civilization of the time.

The progress in cryptanalysis due to Arab scholars laid the foundation for the great evolution in cryptography that occurred during the Renaissance, with Leon Battista Alberti as its prominent representative. From this point on, information security was no longer seen only for military purposes, but its importance for commercial and economic purposes gradually became increasingly evident. This provided a strong impetus for the evolution of cryptography towards a system that is protected from any possible attack.

3.3 Vernam, Shannon, and Absolute Secrecy

The first unbreakable code, the Vernam cipher, was invented in 1917, during World War I, by Gilbert Vernam, an American engineer, although the mathematical proof of its unbreakability was obtained only more than thirty years later by Claude Shannon, the father of modern information theory. The code played a crucial role in the secrecy of communications during the war.

The Vernam protocol uses the following steps:

(1) The message is written as a binary sequence of 0s and 1s (see an example below).
(2) The secret key is a binary sequence of 0s and 1s, completely random, of the same length as the plaintext.
(3) The ciphertext is obtained by adding the key to the message, bit by bit, modulo 2. Modulo 2 addition is such that $0+0=0$, $0+1=1+0=1$, $1+1=0$. For example:

message:	0 0 1 0 1 0 0 1 1
secret key:	1 0 0 1 1 1 0 1 0
ciphertext:	1 0 1 1 0 1 0 0 1

(4) To decrypt the message, it is sufficient to add (modulo 2) the secret key to the ciphertext, bit by bit.

To transform a message into a sequence of numbers, one can, for example, use the Baudot code, devised in 1874 by the French engineer Emile Baudot for transmitting text messages through the telegraph network. Each character is encoded using a sequence of 5 bits, for example the letter A with 11000, the letter E with 10000, the space separating two words with 00000. The possible sequences of 5 bits are 32 in total (00000, 00001, 00010, 00011, ..., 11111), sufficient to encode the 26 letters of the English alphabet, plus the space between words and instructions such as line feed, carriage return, etc. Table 3.1 shows the binary encoding of the phrase "alea iacta est" and its encryption through bit-by-bit modulo 2 addition with a secret key.

The Vernam code is also known as the One-Time Pad, literally "one-time notebook", since the secret key, which must be as long as the message, cannot be reused. In fact, a spy who intercepted two encrypted messages with the same key could add them together (always modulo 2). And this sum would be equal to the sum of the two unencrypted messages, the knowledge of which would allow obtaining information about the original messages.

Table 3.1. An example of binary encoding.

	Message	Secret Key	Ciphertext
a	11000	01101	10101
l	01001	01100	00101
e	10000	01110	11110
a	11000	01010	10010
	00000	10001	10001
i	01100	11110	10010
a	11000	11010	00010
c	01110	10010	11100
t	00001	01101	01100
a	11000	10111	01111
	00000	10000	10000
e	10000	11100	01100
s	10100	10101	00001
t	00001	11010	11011

Although it is a cryptographic system secure against cryptanalytic attacks, the Vernam code has drawbacks. First, it requires a very long key: the length grows with the length of the message. But the main weakness, even if at first glance it may seem strange, lies in the fact that generating long random sequences is not a trivial problem. Moreover, it is not trivial to determine whether a sequence of numbers is random or not. This is a problem that has been studied by mathematicians, who have invented algorithms to generate random sequences or, more precisely, pseudo-random sequences. However, these, if very long, always contain some regularity, which can put the security of a cryptographic protocol at risk.

It is interesting to note that quantum mechanics has also been used for generating random numbers. In fact, schemes for generating random numbers that exploit the intrinsic randomness of the measurement process in quantum mechanics have been studied. Furthermore, devices for generating random numbers based, for example, on single-photon detectors are now commercially available. They represent an important development of practical interest in current quantum technologies.

This example leads us to reflect on how scientific progress in a given field can serve to solve problems in completely different fields, and also makes us understand why it is important to support free research not aimed at a predetermined objective. In fact, it is not possible to determine a priori what things research must discover. It would be useful if this aspect were taken into account by the political class of a country and by the bodies that decide research funding.

Due to the difficulty and cost of providing new random keys for each message to be transmitted, the Vernam cipher is today used mainly for important diplomatic-military communications. It was used, for example, during the Cold War by Eastern secret services and for the red telephone between Washington and Moscow (in reality, it was not a telephone, but two teletypewriters, connected via a cable, mostly underwater, which were used to exchange encrypted written messages).

Then there remains the problem of the secure distribution of the secret key. How can we be sure that a spy has not intercepted the envelope (or any other means of communication) containing the key and made a copy of it to then use it at the appropriate time to decrypt a secret message? As we will see later, quantum mechanics allows solving this problem.

3.4 Enigma: The Invincible Machine

In the period between the two world wars, the difficulties encountered with the Vernam code stimulated the construction of polyalphabetic cipher machines. The most famous machine is Enigma, invented by the German engineer Arthur Scherbius around 1920. Enigma was an electromechanical machine, originally developed for secret commercial communications, but was later adopted by the German armed forces for military purposes during World War II.

The operation of Enigma is very complex but, in simple terms, can be outlined as follows: a keyboard containing the 26 characters of the English alphabet was used to enter the letters of the plaintext. The keyboard was connected to a rotor, i.e., a disk on which the characters of the alphabet were engraved. In this way, a simple monoalphabetic substitution encryption could be established, which, however, as we have seen, can be easily broken. The ingenuity of Enigma lay in the fact that each time a key was pressed, the rotor rotated, changing the correspondence between the characters of the keyboard and those of the rotor. More precisely, the rotor automatically rotated by 1/26 of a turn after the encryption of each letter, allowing the use of 26 cipher alphabets. In this way, the same letter of the alphabet, in the same ciphertext, was encrypted differently. Essentially, Enigma performed a polyalphabetic substitution encryption, which was thus very complex and difficult to decrypt. Later, to increase the degree of security of the machine, Scherbius added two more rotors that rotated according to different rules. In this way, $26 * 26 * 26 = 17576$ different cipher alphabets were used. The key for encryption and decryption consisted of the initial configuration of the rotors, which was different every day. This configuration was transmitted to field units through daily booklets and allowed the recipient of the message to know how to initialize the rotors to decrypt it correctly.

Enigma went down in history as being considered unassailable from its birth. However, towards the end of 1932, this encryption was broken by three Polish mathematicians: Marian Rejewski, Henryk Zygalski, and Jerzy Różycki. On the other hand, between 1938 and 1939, the Germans changed the encryption rules and increased the number of rotors from 3 to 5. The Polish method became inapplicable, and Enigma became unassailable again.

In September 1939, Great Britain declared war on Germany, and one of the first measures of the British government was to strengthen the cryptanalysis center at Bletchley Park, also known as Station X. Located 80 kilometers northwest of London, it had been purchased in May 1938 by Admiral Sir Hugh Sinclair, head of the Secret Intelligence Service (or MI6). The government's commitment was significant, so much so that in January 1945, at the peak of the decryption effort, around nine thousand employees worked at Bletchley Park, including numerous mathematicians, among whom Alan Turing, one of the fathers of computing and artificial intelligence, whom we already met in the second chapter, but also people with diverse cultural backgrounds, from linguists to chess champions. The Daily Telegraph newspaper was even asked to organize a crossword puzzle competition, after which the most promising participants were secretly contacted with an offer of "a special job".

Five weeks before the outbreak of the war, the Biuro Szyfrów (section of Polish intelligence dealing with code analysis) informed the British of the results obtained in breaking the Enigma code. Alan Turing, using the Polish information, succeeded in the feat of decrypting messages encoded with Enigma. The work carried out at Bletchley Park was important for defeating the German Navy's U-boats in the Battle of the Atlantic and, in North Africa, against the Germans led by General Rommel. Before the D-Day landings in June 1944, the Allies already knew the position of 56 out of 58 German divisions on the Western Front. When Italy entered the war in 1940, it adopted an improved version of the Enigma machine. However, the messages of the Italian Navy were decrypted before the Battle of Cape Matapan (the southernmost point of mainland Greece) in March 1941, which favored the British victory. At Cape Matapan, in fact, Admiral Iachino had decided to deploy a significant part of the Italian fleet.

It is worth noting that, in some cases, the information decrypted by the British was not even exploited, so as not to reveal that the British were able to decrypt the messages. The alleged decision not to intervene to save people, i.e., to sacrifice them rather than let the Germans know that Enigma had been broken, has been the subject of heated debates, such as, for example, in the case of the bombing of Coventry, in which more than twelve hundred people lost their lives.

Churchill is said to have stated that Turing was the man who made the greatest contribution to the Allies' victory over Germany. However, in 1952, Turing was convicted of homosexuality and accepted chemical castration to avoid prison. He committed suicide in 1954, and official apologies from British Prime Minister Gordon Brown came only in 2009. A posthumous pardon was granted by Queen Elizabeth II in 2013.

It is also interesting that the construction of one of the first programmable electronic computers in history, Colossus, took place in the United Kingdom during World War II to break the messages of the Lorenz cipher machine, used by the Nazis as a complement to Enigma to protect communications between Hitler and his generals. The continuous challenge of human ingenuity, on the one hand to develop increasingly secure cryptographic codes and on the other to break them, has therefore also been fundamental for the development of computers.

3.5 Secure Communications Today and Prime Numbers

Currently, for common operations such as withdrawing money from an ATM or making online purchases with credit cards, cryptographic techniques called public-key cryptography are used. They are of a completely different type from those discussed so far. The idea of using a public key to encrypt a message represented a step forward in the evolution of cryptography, as it solved the problem of the secure transmission of the secret key. This brilliant invention is due to Whitfield Diffie and Martin Hellman. For simplicity, we focus here on the RSA encryption scheme, introduced by Ronald Rivest, Adi Shamir, and Leonard Adleman in 1977. In reality, according to the British government, public-key cryptography was originally invented at the Government Communications Headquarters (GCHQ) in Cheltenham, but the discovery remained classified until 1997. If we think about it, this is not surprising, since cryptography is a particular science that is not interested in disseminating the results obtained but, on the contrary, seeks to minimize the information available to potential adversaries.

In RSA public-key cryptography, the two communicating parties, who, as usual, we will call Alice and Bob, do not exchange any secret key. One of them, for example Bob, makes a key public (the public key), which Alice uses to encrypt the message. Unlike the Vernam code, there is therefore no problem of securely transmitting a secret key, exposing oneself to the possibility that it may be intercepted. Naturally, the message cannot be decrypted with the public key, but only with another key (the private key), which is possessed only by Bob.

Essentially, there are two types of keys: the public key, i.e., publicly available, which only allows encrypting the message, and the private key, which is known only to the person receiving the message. To make an analogy, it is as if for each message Bob prepares a safe, initially left open, with two locks, each with its own key. One key is public, and Alice uses it to close the safe once the message is inserted. At that point, only Bob can open the safe and read it using his private key.

The main problem of this procedure is how to create two keys such that it is not possible to derive the private key from the public key. Rivest, Shamir, and Adleman succeeded in solving this by using prime numbers. Essentially, given two prime numbers, it is easy to calculate their product. For example, 53 and 139 are two prime numbers whose product is 7367. It is much more difficult, given an integer, to find the prime numbers that, when multiplied together, give that number. In the previous example, given the number 7367, it is not simple in terms of computation time to reconstruct its prime factors 53 and 139. The security of the RSA protocol is based precisely on this. In fact, with the computers and algorithms available today, it is practically impossible to factorize a number longer than 500 decimal digits.

Currently available cryptographic programs use 2048-bit keys (309 decimal digits) which are practically unassailable. The US National Institute of Standards and Technology (NIST) recommends a minimum key size of 2048 bits, but it is considered useful to use 4096-bit keys in situations where data must be protected from attackers who may have state-of-the-art computational resources. The European Union Agency for Cybersecurity, in one of its latest reports on cryptographic issues, also recommends the adoption of 3072-bit keys.

We now show with an example how the RSA algorithm works (the reader who finds this paragraph boring can safely skip it without prejudice to the understanding of the rest of the book). Bob generates two distinct prime numbers p and q, for example $p = 5$ and $q = 11$. We call $N = p*q = 55$ their product. Bob then calculates what is called the "Euler's totient function" of the two numbers, $z = (p - 1) * (q - 1) = 40$, and chooses an integer e that is coprime with z, i.e., the greatest common divisor of z and e is 1. In our case, we can take $e = 3$. Furthermore, Bob calculates the number d as the inverse of e modulo z; in our case $d = 27$, since $d * e = 27 * 3 = 81$, and modulo $z = 40$ this equals 1 (if we divide 81 by 40, the remainder is 1). At this point, Bob makes the pair of numbers (e, N) public, while the number d remains secret, known only to him. The pair (e, N) constitutes the public key with which anyone can send encrypted messages to Bob. A text or an image can be translated into a sequence of natural numbers, for example we have previously seen how to encode the text "alea iacta est". Here we have to encrypt the message by first dividing it into blocks, each of which can be encoded using an integer smaller than N. The letters of the English alphabet are $26 < N = 55$, while the possible sequences of two letters ("al", "ea", ...) are much more numerous than N. Therefore, a block of Alice's message, in our example where the number N is small, consists of a single letter. If a letter is, for example, translated into the number $m = 7$ (for example "a", which is the first letter of the message), Alice encrypts it by calculating the number $c = m^e$ modulo N, so $c = 7^3 = 343$ modulo 55, which equals 13 (the remainder of the division of 343 by 55). The number $c = 13$ is then sent by Alice to Bob via a public channel, for example by publishing it on a website. However, to decrypt c, the private key d that only Bob possesses is needed. It can in fact be proven that $m = c^d$ modulo N; in our case, 13^{27} modulo $55 = 7$. Thus, Bob is able to decrypt the message using his private key and find the number $m = 7$ which corresponds to the letter "a". The strength of the RSA algorithm is that to calculate the private key d, it is not enough to know N, but its prime factors are also needed, and these cannot be calculated efficiently with a classical computer.

The advent of quantum computers, through the use of Shor's algorithm to find the prime factors of an integer, could break the public-key cryptographic codes currently in use. It is very likely not an imminent event, but it is something to be taken into account in the case of information that we want to remain secret for many years. The practical impact of breaking RSA codes with a quantum computer would be significant and would have disruptive consequences in the field of cybersecurity. Since many cryptographic protocols use RSA, their breaking could compromise the confidentiality of communications, data integrity, and authentication in various contexts, including online services, financial transactions, and more.

In this regard, it is significant that the US administration has enacted a law implementing the Quantum Computing Cybersecurity Preparedness Act, which mandates accelerating the migration to post-quantum cryptography as soon as NIST finalizes the appropriate algorithms. Post-quantum cryptography aims to develop public-key cryptographic techniques, which are not yet sufficiently studied, that are not vulnerable to a quantum computer implementing Shor's algorithm and that are at the same time secure against attack by a classical computer. These efforts aim to ensure the security of communications even in a future where quantum computers capable of breaking the RSA protocol become a reality.

3.6 Secure Quantum Communications: What Future?

So far, we have talked about the evolution of cryptographic techniques from ancient times to today. The important question concerns the future. In an increasingly digitalized and interconnected world, information will play an increasingly significant role, permeating various aspects of daily life, the economy, the political sphere, and culture. Various industries, including information technology, communications, finance, and marketing, make extensive use of information management to achieve their objectives. In fact, many of the technological innovations of recent decades have emerged from the management and transmission of information. The Internet, artificial intelligence, and other emerging technologies are examples of how information plays a crucial role in transforming society. Similarly, the massive collection of data, big data, is a tool used in many business and scientific activities. The reason is that data analysis provides useful indications for making decisions in various sectors, from scientific research to business management.

In this context, two critical aspects emerge regarding the potential risks they present. The first concerns the responsible management, access, and critical interpretation of information. In particular, the spread of false or misleading news requires critical attention from the public. The second aspect relates to information security, which has become a matter of crucial importance. The protection of personal data, the prevention of cyberattacks, and the management of the security of computer systems are themes that increasingly pervasively affect our society. This underscores the importance of the topic addressed in this chapter, especially with regard to future developments.

To understand the crucial turning point that is about to occur in this field, let us start with the following reflection: how does information move from one city to another, from one continent to another? Who transports it? From the horseback messengers of ancient times, we have moved to electromagnetic waves that allow us to see images on the television screen, talk on mobile phones, communicate via the Internet, etc. Essentially, all information, of any kind, from remote times to the most modern technologies, has always traveled and travels on a classical medium. That is, the subject that transports the information moves in accordance with the laws of classical mechanics. Despite all the more sophisticated systems designed to protect the secrecy of information that we have just discussed, there is an insurmountable difficulty: it is impossible to know with absolute certainty if a spy has obtained a secret key or has intercepted the message, copied it, and forwarded it to the recipient. The reason is that classical information can be copied without modifying the original message. In fact, information must necessarily be encoded in a physical system (a piece of paper, radio signals, etc.), whose properties can, in principle, be measured "passively". In other words, the modifications induced by a spy reading the message can be minimized to the point of becoming practically irrelevant.

This difficulty could be overcome if it were possible to make information travel not on a classical medium but on a quantum one: not on classical electromagnetic waves but on photons. This possibility, which will revolutionize the world of information, is in an advanced stage of study and in some cases is already a reality. This is what we will talk about in the next paragraph.

3.7 Information Travels on Quanta

In quantum mechanics, as we saw in the first chapter, the measurement process generally disturbs the system. This is a consequence of Heisenberg's uncertainty principle, according to which we cannot simultaneously measure, with arbitrary precision, the position and velocity of a particle. However, Heisenberg's principle is much more general and also applies to qubits. If a qubit is measured, its state is inevitably changed. As we will see later, this inherently quantum property allows detecting intrusions: Alice and Bob can thus discover if a spy has tried to obtain a secret key transmitted via a sequence of qubits.

Based on this opportunity offered by quantum mechanics, and while waiting for future technologies to allow the large-scale transmission of information on a quantum medium, methods for protecting the secrecy of information have been developed for about 40 years. Thus, a new field of research was born: quantum cryptography, which studies how to transmit a secret key in an inherently secure way, leveraging the peculiarities of quantum mechanics.

The first quantum cryptographic protocol was discovered by Charles Bennett and Gilles Brassard in 1984. In this protocol, known as BB84, the goal is to securely transmit a secret key. For this, after writing the key as a sequence of 0s and 1s, two different "alphabets" are used to encode the bits. In the first case (z alphabet), we represent the value of the bit through the 0 and 1 states of a qubit, oriented vertically, that is 0 corresponds to the North Pole of the globe and 1 corresponds to the South Pole, as shown in Figure 3.4. In the second case (x alphabet), the 0 or 1 values of the classical bit are represented by two states of the qubit oriented horizontally, corresponding to two antipodal points on the equator of the globe (just as Quito, the capital of Ecuador, is roughly antipodal to the city of Pekanbaru, on the island of Sumatra, Indonesia). We call N and S (North and South) the two states of the z alphabet; W and E (West and East) those of the x alphabet. It should be noted that, due to Heisenberg's principle, we cannot simultaneously measure with certainty whether the state is North or South and whether it is West or East. The reason is the same as we explained in the first chapter: given a particle, we cannot simultaneously measure its position and velocity.

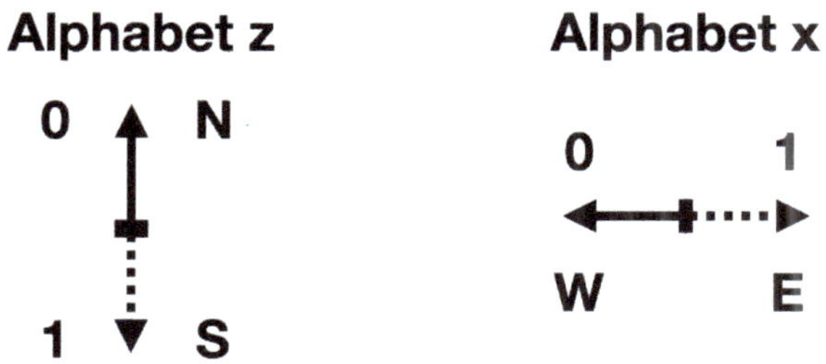

Fig. 3.4. Graphical visualization of the two alphabets in quantum cryptography.

The BB84 protocol works as follows:

(1) Alice generates a random binary sequence of 0s and 1s.
(2) Alice encodes each bit (value 0 or 1) of the key she wants to transmit by randomly choosing between the x alphabet and the z alphabet,

for example by flipping a coin (if, for example, the coin lands heads, Alice chooses the z alphabet; if it lands tails, Alice chooses the x alphabet). Thus, to encode 0 with the z alphabet, Alice prepares a qubit in the N state; to encode 1 using the x alphabet, she prepares a qubit in the E state.

(3) The sequence of qubits thus obtained is sent to Bob. More precisely, physical objects are sent (in practice, photons, that is "particles of light" that form electromagnetic radiation), each prepared in a state corresponding to one of the four possibilities selected above (N, S, W, E).

(4) For each received qubit, Bob also randomly chooses which alphabet to use for the measurement; if he chooses the z alphabet, the possible result he will obtain will be N or S; if he chooses the x alphabet, the possible result will be W or E. If Bob has used the same alphabet as Alice, the value of the bit found by Bob is certainly the value of the bit sent by Alice. Instead, every time Bob chooses a different alphabet, the result of his measurement is completely random. For example, if Alice has encoded the bit 0 using the z alphabet and Bob measures the received qubit with the x alphabet, he will find W or E with equal probability, that is 0 or 1. In other words, the value measured by Bob has no relation to that of the bit encoded by Alice. How can we then arrive at a secure key shared by Alice and Bob? The idea is as follows:

(5) Bob and Alice communicate over a classical public channel which alphabet they used to encode or measure each qubit. Naturally, they only say the alphabet used, not the value obtained.

(6) Alice and Bob then discard the data corresponding to the cases where they used different alphabets, and the remaining data, which they share, constitutes the secret key, known only to them. Since both Alice and Bob randomly choose the alphabets, it is clear that Bob will have chosen the same alphabet as Alice only half the time on average.

Table 3.2 shows an example of using the BB84 protocol to generate the secret key.

The essential point of the protocol is that, due to Heisenberg's principle, if there were an intrusion by a spy, whom we will call Eve, who tried to measure a qubit, this would inevitably modify the state of the system. In fact, if both Alice and Eve randomly choose between the x alphabet and the z alphabet for each bit, half the time their alphabets do not match.

Table 3.2. An example of the BB84 protocol.

Random bit sequence	1	0	0	0	1	1	0	1	0	1
Alice's alphabet	x	z	x	z	x	x	x	z	z	x
Alice's encoding	E	N	W	N	E	E	W	S	N	E
Bob's alphabet	x	z	x	x	z	x	z	x	z	z
Secret key	1	0	0			1			0	

And if they do not match, Eve's measurement, according to the laws of quantum mechanics, inevitably disturbs the system, and this introduces a mismatch in the bits of the key shared by Alice and Bob. The latter can thus easily discover if there has been an intrusion. It is sufficient that, using a public channel, they compare the values of the bits for part of the key (obviously these bits are then sacrificed and no longer used). In this way, they can easily verify if there are mismatches that indicate the action of a spy. In this case, they discard the obtained key and generate a new secret key from scratch using the same BB84 procedure. Once it is confirmed that there has been no intrusion, they can proceed to encode the messages using the Vernam cipher. Quantum mechanics, unlike classical mechanics, thus allows the detection of intrusions: Alice and Bob can discover if Eve is intercepting their communications or not.

At this point, the reader might raise an objection: Eve can intercept the qubits sent by Alice, make a copy of each of them, and finally measure the copied qubits without touching the original ones, so as not to leave any trace of the intrusion. The answer is that this is not possible. In quantum mechanics, it is impossible to make clones. This is the so-called no-cloning theorem, according to which it is not possible to copy a generic quantum state. Photocopiers exist in the classical world, but there is a fundamental difference here: a bit can only be at the North Pole or the South Pole, while a qubit can be anywhere on our globe. Making a copy, an opportunity we take for granted based on our daily experience, is possible for a classical bit but not for a qubit.

In closing this paragraph, we also mention another quantum cryptographic protocol, known as E91, proposed by Artur Ekert in 1991, and potentially interesting for the preservation of secret keys. Suppose that, once the secret key has been established, Alice and Bob want to keep it in their safes until they need it. However, the key is a sequence of classical bits and, in principle, can be copied. It can be very difficult for a spy to open a safe, but it is possible. No fundamental reason excludes this possibility. However, if Alice and Bob were able to share and preserve entangled states, they could wait to establish the secret key until it is necessary to transmit a message. In fact, for an entangled state, the alphabet used for the transmission of each bit of the key remains undetermined until the state is measured. Similar to the EPR paradox described in the first chapter, it is as if Alice and Bob each keep half of a stamp, but it is only with Alice's measurement that it is fixed whether the stamp was cut horizontally (x alphabet) or vertically (z alphabet). According to this procedure, Alice and Bob possess entangled qubits but not even they know the key. This will be generated only when they decide to measure their qubits, and this key will be that generated by the results of the measurement that each will perform in their own laboratory. Obviously, the key thus obtained will be known only to them.

Naturally, the implementation of this idea is hindered by the fact that it would be necessary to be able to protect pairs of entangled states from disturbances due to interactions with the surrounding environment for long periods of time, until a decision is made to generate a key. This possibility is, for now, only theoretical, as it is beyond the reach of current technology. Nevertheless, there remains the conceptual interest related to the possibility of sharing a "potential" secret key while leaving it undetermined and generating it only when necessary.

3.8 Quantum Cryptography Today: From Optical Fiber to Satellite

In the previous section, we saw some clever protocols that allow, in principle, encoding and transmitting information via qubits. However, as we have emphasized several times, quantum objects are very delicate, and even the slightest interaction with the environment destroys their fundamental properties. Will it ever be possible to practically translate these protocols into everyday life and successfully realize an easily accessible technology that allows secure quantum communication? The stakes are so high that a great effort is being made in this direction in many laboratories around the world, and the commercial application of quantum cryptography is already in the experimental phase. Some companies are developing quantum cryptography methods for various applications, in particular for secure communications in financial transactions and intergovernmental communications. This is possible thanks to enormous progress both in optical fiber technology and in the field of free-space optical communication. In both these technologies, information (the transmitted secret key) is encoded via photons, which, we recall, are real "particles of light" that constitute electromagnetic radiation.

A great advantage of quantum communication through optical fibers is that those already installed for classical communications can be used. This makes it possible to leverage existing infrastructure, avoiding the construction of new networks. Single photons are thus injected into the fiber, each prepared in one of the four states belonging to two alphabets, practically realizing the N, S, W, E states of a qubit. The main disadvantage is the loss of photons during transmission through the optical fiber. In fact, simplifying, an optical fiber can be thought of as two very thin glass tubes inserted one inside the other. The inner tube (core) is made of very pure glass containing a high percentage of silica, while the outer tube (cladding), made of less pure glass, causes the reflection of light inside the core. Light propagates like in a tubular mirror through a series of reflections in the inner tube. It is like having a pipe, but instead of water, light flows through it.

The point is that the inside of the fiber, although made of glass of extremely high purity, does not completely eliminate the absorption of photons by residual impurities present in the glass. Similarly, it is not possible to completely eliminate the leakage of photons when the fiber bends during its path (in practice, when a photon hits the curved surface of the fiber instead of being reflected inward, it escapes to the outside and is therefore no longer usable for communication). As the length of the fiber increases, the probability of photon loss increases significantly. More precisely, it increases exponentially with length, and this constitutes a strong limitation on the distance over which quantum communication can be implemented.

For these reasons, based on current technology, it appears prohibitive to use fibers for quantum communication over distances greater than a few hundred kilometers. In principle, quantum repeaters would overcome this limitation and allow extending quantum communication to arbitrarily long distances. A repeater, in fact, captures a signal and retransmits it. Think, for example, of a common Wi-Fi device present in our homes. If the house has multiple floors, we usually need a repeater on each floor so that the Wi-Fi signal is clearly received everywhere, not just on the floor where the optical fiber arrives.

Similarly, it is possible to realize repeaters in quantum mechanics. Currently, the construction of quantum repeaters is still in its embryonic stage. However, the field is developing rapidly, and the first results are encouraging, such as in the experiment carried out by Ben Lanyon's group in

Innsbruck in 2023, schematized in Figure 3.5. Let us analyze the operation of this repeater because it gives us an idea of how quantum technology can be implemented. Two ions, initially prepared in their ground state (that is, their state of lowest energy), are illuminated by laser light from which they absorb energy and thus are brought to a state of higher energy. Subsequently, the two ions decay to the low energy state. They tend, in fact, to naturally return to the state of lowest energy, just as a car, if we do not brake, would roll down a hill without using the engine. In the decay process, each ion emits a photon. In this way, two entangled ion-photon pairs are created. The photons are then sent along 25-kilometer-long optical fibers, one towards node A and the other towards node B. At this point, it is possible to intervene on the state of the two ions, and it is possible, through an appropriate protocol, to transform the entanglement of the ion-photon pairs into photon-photon entanglement, with the two photons being 50 kilometers apart. In this way, it has been possible to transform two entangled connections, each 25 kilometers long, into a single entangled connection 50 kilometers long. Once nodes A and B share entangled states, it is possible to use quantum cryptographic protocols such as the E91 protocol described earlier. While the single repeater just described is what we currently know how to realize, it seems technologically plausible to concatenate repeaters to reach distances of a few hundred kilometers.

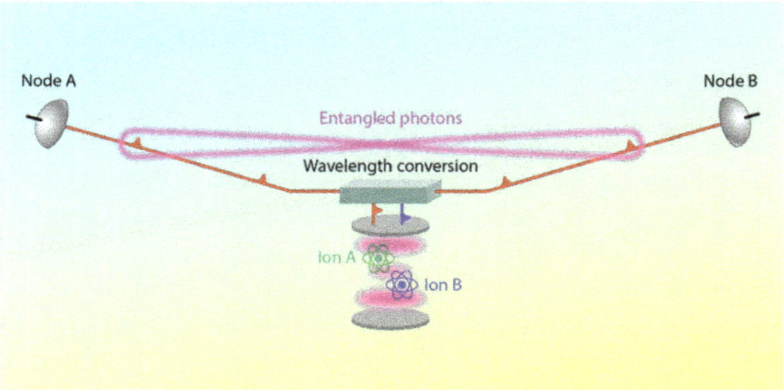

Fig. 3.5. Quantum repeater scheme (Figure taken from `https://physics.aps.org/articles/v16/84`). The entanglement between ion A-photon sent to node A (25 kilometers away) and between ion B-photon sent to node B (25 kilometers away) is converted into entanglement between the two photons (50 kilometers apart).

A completely different technique consists of realizing quantum cryptography in free space, i.e., in the absence of fibers, as has been demonstrated in many experiments. For example, the BB84 protocol was realized in 2007 over a distance of 144 kilometers between La Palma and Tenerife. In this experiment, a source of entangled photons was installed in La Palma, at an altitude of 2400 meters, and one of the photons of each entangled pair was sent to Tenerife and received there by a telescope, also located at an altitude of 2400 meters. Direct connections over significantly longer distances are not possible on the Earth's surface, first of all because they are hindered by the curvature of the Earth (to see the top of a mountain located at a point B, from a point A at sea level and 144 kilometers away from B, the mountain must be at least 1700 meters high).

These limitations can be overcome through the development of quantum technologies in space using satellites (as schematized in Figure 3.6). To create a global-scale quantum communication network, a satellite nicknamed Micius or Mozi has been used, in honor of an ancient Chinese philosopher whose school of thought, opposed to war, advocated universal love and mutual assistance among all people. Micius was also the first to describe the physical principle underlying the camera obscura. Through this satellite, it was possible in 2017 to distribute pairs of entangled photons to two ground stations (Delingha and Lijiang in China), separated by 1203 kilometers. Chinese scientists, in 2021, created the world's first integrated quantum communication network, combining ground-based optical fibers (to connect various banks over limited distances) with two Earth-satellite links to achieve the distribution of quantum keys over long distances, up to 4600 kilometers. China is at the forefront in the development of these technologies. However, in Europe too, a consortium of companies, with the support of the European Space Agency (ESA) and the European Commission, is designing its own satellite system to develop a quantum network for the secure transmission of data. In this regard, the launch of a satellite (Eagle-1) is planned in late 2026.

It is worth emphasizing that, in addition to the practical interest of creating a global-scale quantum cryptographic network in the future, long-distance quantum communication via satellites is also important for other fundamental reasons. In fact, it will be possible to study the effects of gravitational fields on entanglement, and this could provide useful indications on the still elusive link between gravity and quantum mechanics.

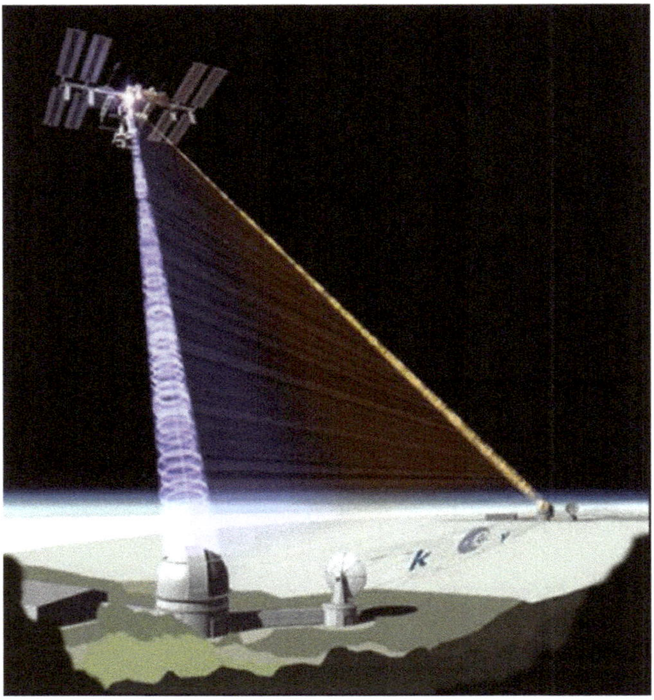

Fig. 3.6. Schematic illustration of entanglement distribution via satellite. Pairs of entangled photons are distributed to two locations on Earth, separated by thousands of kilometers. Figure taken from `https://www.newscientist.com/article/dn14113-space-station-to-be-quantumcommunicationshub/`

3.9 Quantum Teleportation: Science Fiction Becomes Science

Beam me up, Scotty!

James Tiberius Kirk in Star Trek

Among the devices imagined in science fiction works, a particularly dear place to enthusiasts of the genre is certainly occupied by teleportation, made famous by the television series Star Trek, which has captured the imagination of many people, becoming an iconic theme in the world of speculative fiction. Teleportation, in fact, as described in science fiction, would allow the instantaneous transfer of a body (living or not) from one place to another. However, in this sense, teleportation is in conflict with the theory of relativity, which does not allow transfer speeds greater than that of light. Light, in fact, has a finite speed, however high (300,000 kilometers per second), so, for example, it takes an average of about 13 minutes to go from Earth to Mars. Therefore, instantaneous teleportation is not achievable. On the other hand, quantum teleportation of information on subatomic particles is a reality in the laboratory. Moreover, if technologies for large-scale quantum teleportation were ever developed, numerous applications could emerge, mainly related to quantum communication and quantum computing.

Let us clarify this point. Quantum teleportation does not translate into the instantaneous transmission of material objects from one point to another but into the transmission of information, i.e., a quantum state from one particle to another, without the physical transport of the objects themselves as in Star Trek. In this sense, teleportation is a real and fascinating phenomenon, confirmed by many experimental realizations. Furthermore, there is no contradiction with the theory of relativity, since teleportation occurs at the speed at which information travels and thus not exceeding that of light.

Let us take an example. Suppose Alice and Bob both possess a qubit: Alice has a two-level atom, which can be in a superposition of the 0 and 1 states, and Bob has a photon (which can also be in a superposition of two states). With teleportation, the state of Alice's atom becomes the state of Bob's photon, but Alice continues to have her atom and Bob his photon. As we can see, there is no transfer of matter, but of information, i.e., the information that describes the state of a qubit. As Bennett, Brassard, Crépeau, Jozsa, Peres, and Wootters wrote in their 1993 article that introduced the concept of quantum teleportation, "an unknown quantum state can be disassembled and subsequently reconstructed" i.e., whatever the state of Alice's qubit, at the end of the teleportation, it becomes the new state of Bob's qubit. At the same time, Alice's qubit is no longer in its previous state but in a different state. With teleportation, therefore, a physical system is not transported but the state of the system: the state of the qubit has been transferred from Alice to Bob.

It should be noted that this operation of teleporting an unknown state of a qubit seems at first glance a desperate undertaking, impossible to realize. In fact, we have seen that according to quantum mechanics, a measurement of the qubit disturbs its state in an uncontrollable way. After the measurement, the qubit can be found at the North Pole, corresponding to the value 0 of the measurement result, or at the South Pole, corresponding to the value 1. From this information, we are naturally unable to reconstruct the state, i.e., to understand in which state the qubit was before the measurement. Nevertheless, teleportation is possible.

We give below an idea of how this is possible, without going into technical details. The reader who does not want to follow the description of the protocol used for teleportation can skip this paragraph without prejudice to the reading of the rest of the book. Let us start by saying that we need three qubits: one to teleport, plus a pair of entangled qubits shared by Alice and Bob: one qubit of the pair is physically in Alice's laboratory, the other in Bob's. Essentially, Alice takes the qubit she possesses from the entangled pair and makes it interact with the qubit she wants to teleport to Bob. She performs a measurement on these two interacting qubits and in doing so obtains two classical bits of information (the measurement results, which can be 00, 10, 01, 11). Suppose Alice obtains 10 and transmits this information to Bob via a normal classical channel. The crucial point is that Bob, based on this classical information, is able to determine with certainty which operation to implement on his qubit from the initially entangled pair to reconstruct the original state of Alice's qubit to be teleported (we do not go into details here of how Bob chooses this operation). At the end of this operation, Bob has a qubit that is exactly in the state in which Alice's qubit was initially, without it having traveled from Alice to Bob. Furthermore, the original state was destroyed during the measurement; therefore, at the end, the state of Alice's qubit has been teleported from Alice's qubit to Bob's qubit.

What we have just described makes it clear that teleportation does not allow communicating information faster than light. In fact, Alice must send two bits to allow Bob to reconstruct the state. This information, as mentioned earlier, is transmitted by classical means, at a speed that, in accordance with the theory of relativity, will not exceed that of light.

Teleportation is also fully consistent with the no-cloning theorem we discussed in this chapter. In fact, the quantum state of the qubit is teleported from Alice to Bob, while Alice's original qubit, after teleportation, is at the North Pole (state 0) or the South Pole (state 1), depending on the result of Alice's measurement. Thus, Alice's qubit, which was initially in an unknown quantum state, is not cloned but disappears in Alice's laboratory and reappears in Bob's laboratory.

The first experimental realizations of quantum teleportation were obtained many years ago, in 1997, using photons, by Francesco De Martini's group in Rome and Anton Zeilinger's group in Innsbruck. In the following years, many other teleportation experiments were carried out, encoding the state of a qubit using different particles or physical systems (photons, atoms, electrons, and superconducting circuits). In 2017, teleportation experiments from Earth to satellite using entangled photons were also carried out over distances of up to 1400 kilometers by Jian-Wei Pan's group at the University of Science and Technology of China in Hefei.

As we have already observed, the realization of teleportation similar to that depicted in television programs such as Star Trek certainly stimulates our imagination but remains a chimera. It would in fact be necessary to "disassemble" a body in one place, transmit precise information to the destination, and then reconstruct it using that information. An object or even a person would be scanned and encoded, and the transmission of an enormous amount of information would then allow recreating them at the destination. Unfortunately, the challenges associated with this process are enormous and go far beyond our current capabilities. Despite the fascinating developments in quantum physics, it is really difficult to think that teleportation will ever become a real mode of travel.

However, quantum teleportation can have a significant impact in many sectors. For example, it could become an important tool for transmitting information in quantum computers in the future, connecting a network of quantum processors. It would thus be possible to act locally on individual processors, realizing quantum logic gates, and then make the processors collaborate via teleportation to implement quantum computing algorithms. If we want to go a little further, teleportation could be used to improve synchronization between distant space telescopes, improving the resolution of observations, or for synchronizing measurements of quantum sensors over long distances. These applications are, however, at the moment, at the level of pure conjecture and would require significant advances in quantum technologies.

3.10 Energy, Information, and Teleportation

Energy and information are two fundamental resources that characterize the era in which we live. Both play a central role and are closely linked to our daily lives. It is natural to ask if these two elements, so important, are in some way connected to each other. Would it be possible, for example, to transform information into energy and vice versa? At least at first glance, the answer seems to be negative. We know that in our homes, energy is transformed into heat, and a temperature difference is used to produce energy. The possibility of these transformations was at the root of the Industrial Revolution and the development of modern society. We also know, as Einstein taught us, that mass can be transformed into energy, according to the universally known equation $E = mc^2$, where E is energy, m is mass, and c is the speed of light. But information appears to us as something immaterial, subjective, while when we think of energy, we think of something linked to a physical system such as electrical energy, wind energy, thermal energy, etc. How can a transformation occur at the end of which we find, for example, having less information, i.e., being more ignorant, but having more energy?

To understand, let us go back in time to the second half of the 19th century when, through the work of great scientists such as Ludwig Boltzmann and James Clerk Maxwell, the foundations of statistical mechanics were laid. The discussion of the relationship between information and energy can be traced back to Maxwell's demon paradox (schematized in Figure 3.7), introduced by Maxwell in his famous treatise on heat theory, Theory of Heat, in 1872:

> If we conceive a being whose faculties are so sharpened that he can follow every molecule in its course, such a being [...] would be able to do what is at present impossible to us. For we have seen that the molecules in a vessel full of air at uniform temperature are moving with velocities by no means uniform [...]. Now let us suppose that such a vessel is divided into two portions, A and B, by a division in which there is a small hole, and that a being, who can see the individual molecules, opens and closes this hole, so as to allow only the swifter molecules to pass from A to B, and only the slower ones to pass from B to A. He will thus, without expenditure of work, raise the temperature of B and lower that of A, in contradiction to the second law of thermodynamics.

The famous British physicist Lord Kelvin called this creature "Maxwell's demon" for its ability to subvert the natural course of events. The temperatures in portions A and B of the vessel are ultimately different because temperature is, in the final analysis, a measure of the average energy of the molecules, which in turn is determined by their average velocity. We could thus use this temperature difference to produce work, as happens in all heat engines. We could run all machines without any need for oil, gas, uranium, etc.

The paradox arises from the fact that the demon was able to convert heat provided by a single source at uniform temperature (the vessel)

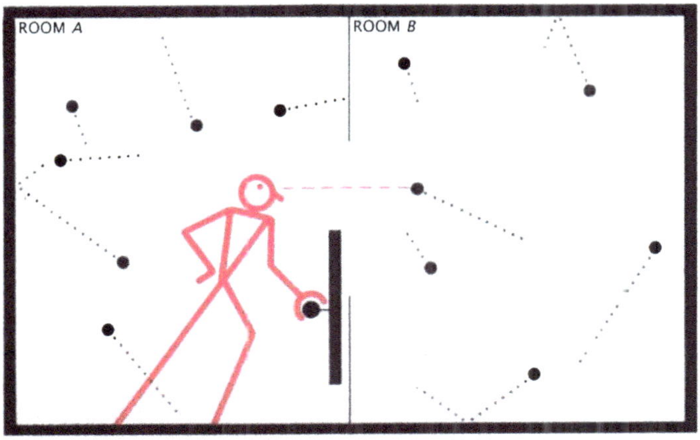

Fig. 3.7. Schematic drawing of Maxwell's demon paradox. The demon, by opening and closing the little door, separates the faster molecules from the slower ones. Figure taken from `https://www.jstor.org/stable/24979551` (Scientific American, 1987).

into work. But this transformation is in contrast with the second law of thermodynamics, and therefore it is impossible. The second law of thermodynamics tells us that not all processes compatible with the principle of conservation of energy occur in nature. If we initially heat only half of our vessel and then allow the molecules to pass through the separating hole, in the end the temperature of the gas will be the same throughout the vessel. The demon performs the reverse process, "orders" the molecules, making the slower ones end up on one side and the faster ones on the other, making the temperatures of the two halves of the vessel different. This process never occurs spontaneously in nature. The principles of thermodynamics play a fundamental role in the description of natural phenomena and are considered to have universal validity. To use Einstein's words, "thermodynamics is the only theory that has universal content, and I am convinced that its conceptual foundations will never be overthrown'.

After Maxwell, for about a hundred years, various attempts were made to reconcile the demon with the laws of thermodynamics. All these attempts proved to be wrong. We want to note that Maxwell was convinced of the validity of the second law of thermodynamics. But the question that arose was more subtle: we cannot violate the second principle because we are not able to do what the demon does; but if there existed an intelligent being capable of doing so, then it would be possible to violate the second principle?

The solution to the paradox was proposed by Charles Bennett and Rolf Landauer based on a principle formulated in 1961 by Landauer himself, a German-born American physicist. This principle states that every time a bit of information is erased, at least a minimum amount of energy is dissipated into the environment. This minimum energy is proportional to the temperature through a universal constant known as Boltzmann's constant. The demon performs measurements on the molecules, and the results of the measurements must be stored in its memory. Since its memory is finite, as all resources for any process in nature are always finite, the demon will eventually have to erase it to free up its "memory cells" and be able to perform new measurements. This erasure operation involves a dissipation of energy, as formalized by Landauer's principle. Essentially, Landauer's principle clarifies that information cannot be erased at no cost: energy is required.

Today, the best supercomputers in terms of energy efficiency can perform up to about 50 billion operations (such as adding or multiplying two numbers) per second and per watt of power consumed. This consumption is about a hundred million times higher than the minimum energy expenditure predicted by Landauer's principle. Current computers are thus far from being energy efficient (incidentally, it is reasonable to assert that they are much less so than the human brain, which has about 85 billion neurons and thousands of synapses per neuron to allow the transmission of information and consumes only 10 watts). To get an idea of the progress made, it is sufficient to think of one of the first electronic computers, ENIAC, put into operation in 1946 in the United States, which occupied 180 square meters and weighed 30 tons (the transistor had not yet been invented at the time, so 18,000 thermionic valves were used for its construction), which consumed about a thousand billion times more per operation than current supercomputers.

To better understand the relationship between information and energy, it is helpful to consider a simple example devised by Charles Bennett. This example demonstrates how information can be used as "fuel" to move a cart. Imagine a cart in an environment with a constant temperature. A tape—composed of a sequence of small boxes, each containing a single molecule—enters the cart (see the upper part of Figure 3.8). Suppose we know the left/right position of each molecule within its box. This information can be used to extract the work needed to move the cart. Here's how:

insert a piston at the center of each box. Knowing the molecule's position tells us how to position the piston (see the lower part of Figure 3.8). If the molecule is on the left side of the box, it will push the piston toward the right wall; if it is on the right side, it will push the piston toward the left wall. In both cases, we know the direction the piston will move, allowing us to extract work proportional to the temperature and the number of boxes. This work can then be used to move the cart. It is important to note that when the tape exits the cart, the molecules can be anywhere inside their respective boxes—the information about their original positions is completely lost, having been "spent" as fuel to move the cart. In the end, we have less information but more energy. Conversely, if we did not know the molecules' positions in advance, we could not extract any useful work. Inserting a piston without this information would be a guess: half the time, the gas would do work on the piston, and half the time, we would have to do work on the gas—resulting in zero net work extracted on average.

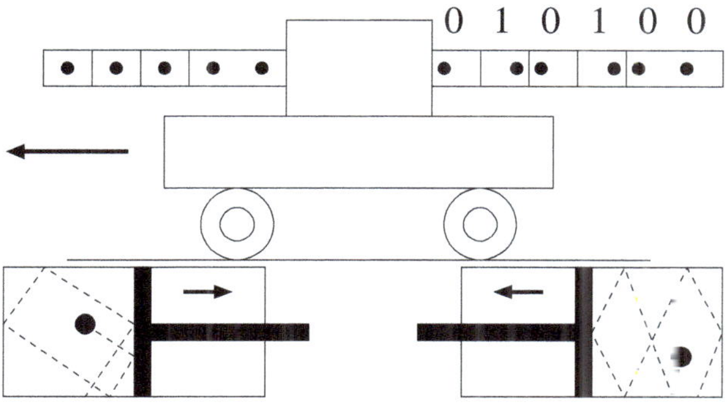

Fig. 3.8. Use of information to produce work. The incoming information (the value 0 or 1 of a bit corresponds to the left or right position of the box for a molecule) is lost at the output.

Notably, Maxwell's demon has been experimentally realized. For example, in 2014, Jukka Pekola's team at Aalto University in Finland created a system analogous to a box containing a single molecule. They used a single electron that could move between two tiny metal islands (each roughly one micron in size). A single-electron transistor (a modern technological version of the demon) was used to track the electron's position. By later erasing this position information, the team was able to extract work, similar to how information was used as fuel to move the cart in the earlier example.

It is too early to say whether these recent experimental advances will lead to new, more heat-efficient computer chips in the future. Even so, it is fascinating that Maxwell's demon paradox—and the subsequent debate about the relationship between information and energy—can now be explored in laboratory settings.

After exploring the deep connection between energy and information, we can take one step further: energy teleportation. Yes, it is possible to transfer not only information but also energy, as if it were "teleported" from one location to another. The secret lies in that remarkable quantum connection known as entanglement: Alice and Bob must first share an entangled state. It is as if they hold the invisible threads of a quantum network, ready to transmit not just data, but also power.

First, we must recall a key consequence of the uncertainty principle: a quantum system cannot settle into a state of perfect rest. Consider a pendulum clock—if not wound regularly, it will eventually stop at its equilibrium position, with the pendulum hanging vertically downward. A quantum pendulum, however, can never come to a complete stop. If it did, we could simultaneously know both its position and velocity—violating Heisenberg's uncertainty principle. This means its energy can never be exactly zero. This is a general rule: the energy of a quantum system can never be completely eliminated, even in its lowest possible energy state (known as the "vacuum state"). Moreover, the uncertainty principle means we cannot know both energy and time with absolute precision. As a result, even in the vacuum state, energy fluctuates over time—with small but non-zero values. These energy fluctuations cannot be used to power a machine, however, because the fluctuations at any given point are completely random. If we tried to connect a battery to a vacuum-state system, half the time the fluctuations would charge the battery, and half the time they would drain it.

Once again, entanglement comes to the rescue. If Alice and Bob share an entangled state (which we can imagine as two entangled qubits, one held by Alice and the other by Bob), the energy fluctuations in Alice's laboratory are correlated with those in Bob's laboratory. Thus, even if the total system is in its lowest energy state, Bob can extract energy in the following way: Alice measures her qubit and sends the measurement result to Bob via a classical communication channel. It can be shown that Bob, using this information, can perform an operation on his qubit to raise it to a higher energy state than its ground state. In essence, Bob leverages Alice's message and their initial shared entanglement to increase his qubit's energy—as if he had "charged" a "quantum battery". The only limitation of this operation is the speed of light, which governs the transmission of Alice's classical message to Bob.

Energy teleportation was first proposed in 2008 by Masahiro Hotta of Tohoku University in Japan. In 2023, Raymond Laflamme's team at the University of Waterloo in Canada experimentally demonstrated it for the first time using nuclear magnetic resonance techniques. Future applications could include providing localized energy to individual qubits in a quantum computer—supplying the power needed to run quantum algorithms. From a conceptual standpoint, it is intriguing that quantum mechanics not only preserves the link between information and energy but also strengthens it, enabling the teleportation of both.

Chapter 4

Sensors of the Future: Perceiving the Impossible with Quantum Technology

What we observe is not nature
itself, but nature exposed to
our method of questioning.

Werner Heisenberg

4.1 Internet of Things

The modern world is increasingly permeated by sensors—they are everywhere and getting smaller all the time. The enormous growth in their numbers and the technological potential offered by sensors integrated into the electronic devices surrounding us have made the vision and development of the Internet of Things (IoT) possible. This fascinating vision is becoming a reality, promising to revolutionize our daily lives with the same disruptive force that the Internet has exerted over the past few decades. Computers, smartwatches, household appliances, vehicles, and more—these devices not only fulfill their designed functions but also act as "living parts" of a broader digital ecosystem. They interact and communicate with other electronic devices within a given environment, such as a building or even an entire city, giving rise to the concepts of smart homes and smart cities alike. Today, this compelling prospect constitutes a market estimated at 300 billion US dollars, with an expected annual growth rate of approximately 20% in the coming years.

Countless examples of this reality already exist: alarms that ring earlier if there is traffic on the commute to work, the automatic delivery of insulin regulated by constant monitoring of glucose levels in diabetic patients, the automated management of household appliances to use them during off-peak consumption hours, smartwatches that display messages or emails instantly while also monitoring athletic performance and vital signs like blood pressure and heart rate, and video doorbells and alarm systems connected to the network that can be remotely controlled via a smartphone. In the automotive and transportation sectors, sensors are essential for vehicles to perceive their surrounding environment, enabling safe autonomous driving. In agriculture, they are used to monitor soil quality, crop growth, and water management. Finally, sensors are fundamental in smart cities for collecting data on traffic management, public lighting, air quality, and more. This vibrant reality animates our existence, weaving our lives more closely with technology, making them increasingly "connected" each day, and enhancing their quality in ways we once could not have imagined.

The growth of the IoT goes hand in hand with the development of artificial intelligence (AI). AI techniques enable the real-time collection and processing of data from moving vehicles to monitor and manage traffic conditions, or to synchronize traffic lights in the event of accidents with casualties to allow ambulances to arrive more quickly. AI can also detect risky driving behaviors, helping to make our roads safer.

Alongside these fascinating prospects come significant security and privacy challenges. For instance, we need to prevent malicious actors from remotely opening our front doors or disabling our alarm systems. Furthermore, sensitive data such as health information requires maximum protection. This is why cybersecurity has become an increasingly pressing issue in contemporary society—without proper security measures, anyone could improperly access various aspects of people's private lives. In a dystopian vision of the future, one might even view the IoT as a tool to restrict individual freedoms by controlling people's activities, contacts, and movements.

Concrete ways in which the IoT could impact individual freedoms include:

(1) *Personal device monitoring*: If everyday objects or wearable devices are equipped with IoT sensors and connections, a person's daily activities or health data could be monitored.
(2) *Geolocation*: If connected objects feature geolocation technology, this can lead to constant monitoring of people's locations, with implications for their freedom of movement and the privacy of their activities.
(3) *Home automation*: IoT-based home automation systems can collect data on individuals' household habits and routines. If this information

falls into the wrong hands, it could be used for unethical purposes such as identity theft or planning criminal acts.

(4) *Shared health data*: Connected medical devices can transmit real-time health data. If this data is not adequately protected, it could be accessible to third parties.

To address these concerns, it is essential to develop appropriate regulations, robust security policies, and empower citizens to control their personal data. Balancing technological innovation with respect for privacy is crucial to ensuring that the IoT contributes to social well-being without compromising individual freedoms.

4.2 Sensors and Scales: The Art of Measurement

From a practical perspective, everyday objects integrated into the digital world through the IoT can acquire and exchange information about their surrounding environment via various types of sensors. The IoT fundamentally relies on the ability to embed sensors into increasingly portable and cost-effective devices. It is therefore clear that, much like the microprocessors in modern computers, the natural path of sensor development is toward creating smaller and more precise sensors. Once again, we might ask: What could happen when sensors are as small as a few atoms? The answer is that the rules of quantum mechanics will begin to play a role—this could be a limitation on one hand, but on the other, it could unlock new possibilities inaccessible to classical sensors. Let us therefore seek to understand what a sensor is and whether the rules of quantum mechanics can enhance its core function: acquiring information about a specific physical quantity.

Readers should not be surprised that we dedicate some space in this book to sensors, as they will increasingly permeate our lives in numerous ways. The evolution toward Industry 4.0 (the so-called Fourth Industrial Revolution) relies heavily on the integration of sensors to improve efficiency and productivity. Their use will continue to expand, contributing to greater efficiency, safety, and quality of life. This is particularly important for Industry 5.0, which represents a significant evolution of Industry 4.0 by placing greater emphasis on human-machine interaction and aiming for optimization that is not only technological but also social and ecological. We therefore believe it is useful to explore how sensors will evolve in the near future through the introduction of quantum technologies.

A sensor can be defined as a device whose state depends on the value of an external physical quantity. The user must be able to read the result of a measurement performed by the device, thereby acquiring information about the physical quantity of interest. This operation is so natural that we perform it daily without realizing it, to interpret various aspects of the environment we live in. Often, however, this process requires sophisticated levels of abstraction and standardization.

Consider when we check the temperature of our home or measure a fever: in both cases, we are using a sensor—a thermometer. Modern electronic thermometers have greatly simplified this process, displaying the temperature on a screen for extreme ease of reading. However, this has not always been the case. Mercury thermometers, used until relatively recently, required manual initialization. The thermometer had to be skillfully shaken to bring the mercury below the first mark before measurement could begin. Subsequently, the thermometer had to be precisely placed on the body, and one had to wait for the mercury to react to the temperature and rise up the column. The final reading—a veritable rite of passage for children—involved a certain skill in handling the device to correctly read the temperature indicated by the mercury column.

This measurement process actually conceals a highly complex procedure: the definition of a scale and a unit of measurement for the quantity being measured. While Celsius degrees (also called centigrade) seem natural to us, it is worth remembering that even today, centuries after this scale was first proposed by the Swedish astronomer Anders Celsius in 1742, there is no single international standard. A significant portion of the world still uses Fahrenheit degrees. The same situation applies to all commonly used units of measurement (such as length, mass, temperature), where the International System (lengths in meters, masses in kilograms, time in seconds, temperatures in Celsius), the centimeter-gram-second (CGS) System, and the Imperial System—still used in the United Kingdom and, with some modifications, in the United States (lengths in inches or feet, masses in ounces or pounds, temperatures in Fahrenheit, etc.)—coexist. For other units of measurement, multiple scales even coexist, typically for historical reasons. This is the case, for example, with the measurement of magnetic fields, where Gauss or Tesla are equivalently used. Complicating matters further, different scales are sometimes used to measure different aspects of the same physical phenomenon: one scale measures the value of the physical quantity, while the other measures its effect on humans or the environment. When we hear news of an earthquake on television, the first piece of data communicated is its intensity. Sometimes this intensity is provided on the Mercalli scale, other times on the Richter scale. These are two distinct scales: the former measures the earthquake's effect on the land, while the latter measures the energy released by the seismic event. The two scales are not directly correlated, as the damage inflicted on the land—and thus the Mercalli scale rating—depends not only on the earthquake's energy but also on soil conditions, the sturdiness of buildings, and so on.

The adoption of universal, shared units of measurement is fundamental to ensuring fairness and precision in trade and commerce. While this concept may seem simple, it has undergone continuous evolution driven by technological innovation. In ancient civilizations, from Mesopotamia and Egypt to Greece and Rome, the cubit was used as a unit of length. This term refers to the length of the forearm, from the elbow to the tip of the middle finger. In the Book of Genesis, it is written that "the length of the ark shall be three hundred cubits, its breadth 50 cubits, and its height 30 cubits"—equivalent to approximately 133 meters in length, 22 meters in width, and 13 meters in height, assuming the Hebrew cubit is 44.45 centimeters long. However, the use of the cubit as a measurement presented significant complications. Its length varied substantially from region to region, ranging between 43 and 50 centimeters, and could also change over time depending on the forearm length of the reigning sovereign.

The absence of uniform measurement standards has also often been a source of fraud in commercial transactions. An example is the so-called Mexican vara, used during Spanish rule in Latin America The vara was a unit of length equivalent to approximately 0.84 meters, used to allocate or sell land. However, its exact value could vary significantly from one region to another. In land sales or concessions, sellers could claim that a vara was longer than it actually was, thereby increasing the apparent area of the land and, consequently, the price obtained.

With the growth of trade and the Industrial Revolution, the need to unify length measurements became increasingly evident. In 1791, the French National Assembly made a decision that could be described as revolutionary: the meter was defined as one ten-millionth of the distance from the Equator to the North Pole, measured along the meridian passing through Paris. Subsequently, precision was further improved with the definition of a "prototype meter"—a platinum-iridium bar with two marks defining its length, kept in Sèvres, near Paris. However, even platinum-iridium is not immune to the passage of time: temperature variations and the loss of even a few atoms through evaporation could alter its length over centuries. The search for an even more reliable measurement led to a completely new definition of the meter. In 1983, the Seventeenth General

Conference on Weights and Measures adopted a definition based on a fundamental principle of nature. The meter is the distance that light travels in vacuum in a precisely defined fraction of a second—1/299,792,458 of a second. This definition leverages a universal constant, the speed of light in vacuum, which, as stated by the theory of relativity, is the same anywhere in the universe. In this way, not only is unprecedented precision in measurement guaranteed, but the unit of measurement also assumes a universal value. Indeed, it does not refer to the circumstances of our terrestrial civilization or the dimensions of our planet, but retains its meaning for all time and all places, whether terrestrial and human or not.

At this point, we cannot resist the temptation to briefly touch on the meaning of a universal constant. It is true that the speed of light is a universal constant—a cornerstone of our knowledge. However, its numerical value depends on the units of measurement used; in this case, meters and seconds. In our universe, there exists an even more fascinating universal constant, as it is free from the influence of any unit of measurement: it is a pure number. This is the famous fine-structure constant, denoted by α, whose numerical value is:

$$\alpha = 2\pi \frac{e^2}{hc} \approx \frac{1}{137.036},$$

where, aside from the factor 2π, it is the ratio of the square of the electron's charge to Planck's constant (h) multiplied by the speed of light (c). This constant is of fundamental importance. If its value were slightly different, our universe would be completely different, and we humans might not exist. The fact that this constant takes on precisely this value, and not another, remains one of the great mysteries of science—a intellectual challenge that continues to stimulate the curiosity and imagination of researchers around the world. Richard Feynman, one of the brightest minds in 20th-century physics, wrote that the reason why this constant has this particular value remains "a mystery, and all good theoretical physicists put this number up on their wall and worry about it."

4.3 The Sensors and the Fundamental Forces of Nature

Let us return to sensors. As we have seen, a sensor is nothing more than a tool for measuring a physical quantity. The measurement process conceals a dialogue (an interaction), often invisible to our eyes, between measuring instruments and the universe we seek to understand. Take the thermometer again, for example: it reveals the temperature of a room—that is, the "measurement"—through interaction with the surrounding air. What we perceive as temperature is actually the energy of air molecules, which move randomly as they collide with the thermometer. But if we observed this scene up close, through a lens capable of transporting us into the microscopic world of particles, we would discover that our concept of "collision" transforms. Atoms and molecules move, guided by electric forces, bouncing off each other without ever truly touching.

However, electric forces are not the only ones existing in nature; on the other hand, a sensor's operation depends on the forces through which it interacts with the external world. So we ask: Beyond the electric force, what other forces are present in our universe? This brings us back to the 17th century, when Newton demonstrated that both the force keeping us anchored to the Earth and the force maintaining celestial bodies in orbit are expressions of the same force: the gravitational force. Later, in the 19th century, Faraday and Ampère showed that electric and magnetic forces are actually the same force, and subsequently, a great genius of physics, James Clerk Maxwell, wove these electric and magnetic forces together, discovering the existence of electromagnetic waves. Today, we know that both light and radio waves are nothing but electromagnetic waves. Finally, in the 20th century, the scene was enriched by two other forces: the weak nuclear force and the strong nuclear force, which explain radioactivity and the "glue" that holds protons together in atomic nuclei, despite their natural repulsion, respectively. Protons, having the same electric charge, tend to repel each other, and the strong nuclear force, opposing this repulsive force, keeps them bound in the nucleus.

Today, we know that all forces present in nature can be reduced to these four fundamental forces: the electromagnetic force between electric charges, the gravitational force between masses, and the weak and strong forces that act within atomic nuclei. These four fundamental forces—electromagnetic, gravitational, weak, and strong—thus orchestrate the universe. While the latter two operate on very small scales, understanding and measuring electromagnetic and gravitational interactions opens the door to a profound comprehension of our world. Sensors, which respond to these forces, are the tools with which we listen to the heart of nature—an essential step forward in the path of modern science and technology.

Once a sensor has modified its state through interaction with the object being measured, the next step is to decipher this change to transform it into the desired information. Let us take the example of temperature again: How can a thermometer, interacting with air molecules, tell us the exact temperature? To do this, a standardized process is adopted, including an initial reference point and a measurement scale. Everyone knows that, according to the Celsius scale, water freezes at 0 degrees and boils at 100 degrees, assuming the thermometer is calibrated at sea level and standard atmospheric pressure. However, this is only one possible convention, and as we have seen, others are equally valid—for example, the Fahrenheit scale.

This same principle of calibration and convention applies to all other sensors, regardless of the quantities being measured, whether pressure, humidity, luminosity, or wind speed. For instance, to determine the concentration of ions in a solution, we leverage the fact that ions, being charged molecules, generate an electric current proportional to their concentration. By measuring this current, we deduce the ionic concentration. Similarly, a photoresistor changes the current passing through it in response to the light intensity it receives: by measuring this current, we trace back to the intensity of the light source. In the same way, ultrasonic sensors calculate the distance to an object by timing the return of sound echoes, while infrared sensors, common in security systems, detect electromagnetic radiation emitted by warm bodies such as that of an intruder, triggering only when the radiation exceeds a certain threshold to avoid false alarms caused by small animals (for example, we do not want the alarm to sound if a cat passes by). In all these cases, the key lies in sensor calibration and the choice of measurement scale. Once these parameters are defined, the sensor can provide precise measurements of the physical quantity in question, enabling the result to be communicated via a simple number and the corresponding unit of measurement.

Since the end of the last century, researchers have begun to theorize and experiment in laboratories with the creation of sensors that obey the laws of quantum mechanics. To fully appreciate the revolutionary potential of these devices, it is fundamental to consider the transformations that quantum mechanics introduces into interaction and measurement processes. We have already seen the substantial benefits it has brought to the fields of computing and information transmission. The question now naturally arises: How can quantum innovation enhance the performance and effectiveness of sensors? We will seek to answer this question in the remainder of this chapter.

4.4 Rabi Oscillations and Single-Qubit Quantum Sensors

Unlike the complex requirements needed to create a quantum computer, the field of quantum sensors is more accessible and versatile, paving the way for a wide range of possibilities and applications. The core of this type of sensor lies in its ability to prepare superpositions of states in accordance with the laws of quantum mechanics, even without necessarily using entanglement—a key ingredient for quantum computing.

Imagine looking beyond what our eyes can see, entering the realm of particles and waves. Since the dawn of quantum mechanics, we have realized that atoms—the fundamental building blocks of matter—behave in an apparently mysterious way when influenced by electromagnetic fields. In the quantum world, they emit and absorb radiation only at specific frequencies. It is as if electrons, in their movement around the nucleus, can only make predetermined quantum jumps from one energy level to another, and they perform these jumps only when they receive the right energy from the electromagnetic wave striking them. This scenario led to Planck's famous formula, which links the energy of an electromagnetic wave to its frequency, and revealed the quantization of atomic energy levels—that is, the utterly counterintuitive fact that not all values are possible for the energy levels of atoms. An electromagnetic wave can induce a quantum jump only when its frequency ν is related to the energy difference E between two atomic energy levels according to Planck's formula $E = h\nu$, where h is Planck's universal constant.

The fact that transitions from one energy level to another occur only at specific frequencies tells us that not all energies are possible for an electron orbiting a nucleus—only certain energy values are. This means that energy levels are "discrete", or "quantized". This phenomenon was fundamental to the development of quantum mechanics, but it also testifies to the fact that atoms can be excellent sensors of electromagnetic fields. For example, we can use them to detect whether a particular frequency of radiation is present in space. If an atom jumps from one energy level to another (in short, "excites") and then releases light as it decays back to its original level, this means that radiation at the "right" frequency to excite it is present (in accordance with the Planck formula just written). Thus, atoms are not only passive spectators of this quantum spectacle but also sensitive detectors of electromagnetic fields.

But that's not all: atoms can also be accurate detectors of static magnetic fields. Thanks to the Zeeman effect, discovered by the Dutch physicist

Pieter Zeeman (Nobel Prize in Physics in 1902, who was also a skilled violinist), we know that in the presence of a static magnetic field, the frequency at which an atom absorbs radiation depends on the intensity of the magnetic field itself. Measuring the frequency of the absorbed radiation therefore allows us to precisely determine the intensity of the magnetic field, opening the door to a new generation of quantum sensors capable of measuring the action of magnetic fields on a single atom. For example, ions trapped using the techniques developed for quantum computers described in Chapter 2 can be used to make accurate measurements of the spatial variation of magnetic field intensity, with a resolution of a few nanometers (one nanometer corresponds to one millionth of a millimeter).

But how do we measure the frequency at which an atom absorbs radiation? This is where Rabi oscillations (named after Isidor Isaac Rabi, Nobel Prize in Physics in 1944) come into play. Imagine an arena where atoms and beams of electromagnetic radiation intersect. This spectacle is at the core of tools ranging from quantum sensors probing the boundaries of science, to magnetic resonance machines scrutinizing the human body, to the fundamental building blocks of quantum computers. The secret to understanding Rabi oscillations lies in how atoms "perceive" the presence of an external electromagnetic field. Imagine having two energy levels, like two notes in a musical score. When light strikes at the right frequency—like a conductor setting the tempo—atoms begin to oscillate between these two "notes" at a rate that depends on the light intensity. These oscillations are the key to deciphering the intensity of the electromagnetic field, transforming atoms into sophisticated quantum sensors. Each time we measure this rate, we enter the realm of probability, where quantum mechanics allows us to predict how often we will find the atom in one note rather than the other. By repeating the experiment numerous times, we can obtain an increasingly precise picture of these oscillations, as if we were tuning a delicate musical instrument. Essentially, by measuring the state of the system at different times, it is possible to reconstruct the oscillation frequency between the two "notes". Since this frequency depends on the intensity of the external electromagnetic field, the quantum sensor ultimately measures this intensity.

However, these oscillations are disturbed by interactions with the unruly crowd of atoms and other particles in the external environment, which tend to dampen them. The smaller this disturbance—known as decoherence—the better we can leverage the quantum realm: the more oscillations we can observe before decoherence dampens them, the better we can measure their frequency.

Maintaining Rabi oscillations, and thus quantum coherence, for increasingly longer periods is a challenge for quantum technologies. This coherence has the potential to revolutionize how we measure, observe, and interact with the world at the subatomic level. Thanks to Rabi oscillations, we can design tools that one day may radically transform how we live, work, and care for our health.

4.5 Illuminating the Invisible: Nuclear Magnetic Resonance

The same fundamental mechanisms underlying quantum sensing are the basis of Nuclear Magnetic Resonance (NMR). Based on Rabi's pioneering insight, this technique—a precious gem in the treasure trove of modern technology—has evolved thanks to the brilliant minds of Felix Bloch and Edward Purcell, who were awarded the Nobel Prize in Physics in 1952 for transforming an abstract concept into an instrument of enormous utility.

Originally explored by chemists to peer into the heart of matter and by physicists in search of secrets hidden in material structures, NMR has since demonstrated remarkable versatility, ranging from the pharmaceutical industry—where it plays a crucial role, for example, in determining the purity and quantity of active ingredients in the development of new drugs—to the depths of the Earth, where it helps uncover the composition of hydrocarbons.

However, it is in the vast ocean of medicine that NMR plays a primary role, offering a window into the hidden world inside our bodies. It can diagnose tumors and many other diseases or abnormalities of internal organs without the need for invasive procedures or exposing the patient to harmful radiation, unlike Computed Tomography (CT), which involves risks associated with high exposure to X-rays.

It is important to emphasize that in medical applications, as in other uses of NMR, the focus is not on the signal derived from a single atom, but on the average signal produced by a certain type of atom (such as hydrogen) present at a specific concentration in human tissue under the action of an appropriate electromagnetic field. NMR uses a very powerful magnet to generate an intense magnetic field—several tens of thousands of times stronger than the Earth's magnetic field. When a patient enters this magnetic field, the nuclei of hydrogen atoms in the body (primarily in water and fatty tissues) align with the direction of the field. Once the nuclei are aligned, the NMR machine sends radio waves (radiofrequency pulses) toward the body. These energetic pulses cause the aligned hydrogen nuclei to temporarily move away from their aligned state. After the radiofrequency pulse is turned off, the nuclei tend to return to their original state, realigning with the magnetic field. In doing so, they emit radio signals, which are measured by dedicated receivers. The process by which nuclei return to their original state occurs at different speeds depending on the type of tissue they are in. The signals emitted by the nuclei are collected and analyzed by a computer. The way in which different tissues release energy and return to their original aligned state allows distinguishing between different types of tissues. Using these data, a computer creates images of the examined area, which can be extremely detailed and prove particularly useful for visualizing the brain, muscles, heart, and tumors without the use of ionizing radiation. While NMR uses a chorus of atoms to map the inner world of the human body, quantum sensors "listen" to a single atomic voice. As we will now see, by leveraging the laws of quantum entanglement, quantum sensors can surpass the limits of their classical predecessors, opening the door to a new realm where the very fabric of reality can be measured with unprecedented clarity.

4.6 The Heisenberg Limit and the Revolution of Quantum Sensors

Imagine having to measure the length of a dining table. With a ruler at hand, in a few seconds we could say it measures, for example, 127 centimeters. In everyday life, an approximate estimate like this is often sufficient, but scientists require much more: they need to know how reliable that measurement is. In other words, it is fundamental to know the error associated with it—that is, how much precision lies behind that number. Numerous factors can influence a measurement: perhaps the ruler was not perfectly aligned, maybe we just took a quick glance and missed that extra millimeter or so, or perhaps a momentary distraction caused us to count incorrectly. When it comes to precision measurements, potential sources of error are as varied as they are subtle—many difficult to control and some even unknown. This leads us to ask: How can we truly be certain of the precision of a measurement?

The solution lies in a scientifically rigorous and universally recognized protocol for estimating errors. This process, familiar to many science students from their first year of university, involves repeating measurements under ideally identical conditions a significant number of times. The underlying idea is that if the errors made during measurements are not systematic (for example, due to a poorly calibrated or improperly used instrument), all unwanted effects influencing the measurement will result in random error—sometimes increasing and sometimes decreasing the result.

In other words, if we measure a table one hundred times, we obtain a list of non-identical results—sometimes 127.1 centimeters, other times 126.7, and so on. From this list of data, how do we reconstruct the measurement result and the measurement error itself? This is where statistics come into play. By repeating the measurement multiple times, we can not only find the average value, which gives us the most accurate idea of the table's length, but also discover how "scattered" the various results are around this average value. This dispersion is known as the standard deviation, and it provides a valuable indication of the typical imprecision of our measurements. In practice, it is a methodology that allows us to find our way through the forest of data, offering a reliable compass to navigate between small variations and apparent contradictions, helping us construct a final measurement and its associated precision.

Essentially, by measuring the table one hundred times, we could, for example, state with certainty that it is 127 centimeters (the value calculated as the arithmetic mean of all repeated measurements), with an uncertainty margin of 0.3 centimeters. The standard deviation is no less important than the mean. In blood tests, for example, we have reference values—a lower and an upper limit—for the tests performed. These values are derived from the distribution of results of that test in the population. Only if we find a value that deviates significantly from the interval between the lower and upper reference values—and thus corresponds to a deviation of several standard deviations from the population mean—should we consider it abnormal.

We are interested in having sensors that provide responses with the highest possible precision. So how can we increase measurement precision? The solution comes from the law of large numbers, one of the fundamental rules of statistics: the more measurements we take, the closer our average approaches the true value of what we are measuring. Moreover, the error in our measurements not only decreases but does so following a precise rule: it decreases proportionally to the square root of the number of times we repeat the measurement. For example, if measuring a table 100 times reveals an error of one millimeter, to refine our precision to one-tenth of a millimeter—that is, improve precision tenfold—we would need to perform the measurement no fewer than 10,000 times! And if we want to achieve precision to one-hundredth of a millimeter? We have to measure one million times. In practice, this means that every time we double the number of measurements, the error is reduced not by half, but only by the square root of two.

It may come as a surprise to learn that this limit to error reduction—known as the shot noise limit—is not due to any shortcoming in our ability to take accurate measurements, but represents a theoretical boundary. It tells us that no matter how much we refine our techniques or how much precision we can embed in our measuring instruments (imagine building an instrument ten times more precise), reducing error through repeated measurements will always follow the same rule: it will only improve proportionally to the square root of the number of measurements performed.

But is there any possibility of exceeding this limit imposed by classical measurements? Once again, quantum mechanics comes to our aid through phenomena like entanglement, which links particles or systems in ways that defy common sense. Through entanglement, the properties of one particle can be immediately connected to those of another, no matter how far apart they are. The strategy here is radically different from the traditional classical approach, which requires repeating a measurement many times with a single sensor to minimize errors. With the quantum approach, we use entanglement to synchronize many sensors so that they operate as a single complex system. This coordinated group of sensors can perform measurements with precision far beyond the limits of classical techniques, opening new frontiers in the collection of accurate and reliable data. In this context, the so-called "Schrödinger's cat states" play a fundamental role. These states take their name from the thought experiment described by Schrödinger, already discussed in Chapter 2, in which a cat is considered both alive and dead until the moment of observation. In the context of sensors, this concept translates to preparing N qubits in a superposition state. A Schrödinger's cat state is given by the superposition of the state in which all qubits are at the "North Pole" (equivalent to the "alive cat") and the state in which all qubits are at the "South Pole" (equivalent to the "dead cat"). These sensors are "entangled" and interact collectively with the object being measured. The fundamental point of this approach is that, instead of performing N separate measurements, the sensors carry out a single collective measurement. The extraordinary net result is that the measurement error does not decrease following the classical square root law of the number of measurements, but decreases proportionally to the total number of sensors involved. This represents an enormous improvement and revolutionizes how we can collect and interpret data.

This significant reduction in error, which decreases proportionally to the number of sensors involved, is known as the Heisenberg limit. It represents the maximum level of precision theoretically achievable in our universe for measurements, and it is obtained by fully leveraging the properties of quantum entanglement. It is a qualitative leap compared to classical methods because, by exploiting quantum correlation between sensors, much more precise information about the object being measured can be obtained.

Entanglement thus elevates quantum sensors to a level of precision and sensitivity with which traditional sensors cannot compete. Consider a theoretical experiment: a classical sensor is ten times more precise than a single quantum sensor. At first glance, it might seem that the classical sensor has an advantage. But entanglement changes the rules of the game. By linking ten quantum sensors in an entangled state, we achieve the precision of the classical sensor. With one hundred entangled sensors, we improve precision tenfold compared to the classical sensor; with one thousand, one hundredfold. This ability to refine precision through entanglement demonstrates truly extraordinary scalability and potential. This provides a significant competitive advantage to quantum sensors, making them potentially revolutionary tools in a wide range of applications—from particle physics to medicine, from astronomy to cybersecurity.

However, the practical realization of large-scale entangled quantum sensors represents an exceptional technological challenge. Creating and maintaining an entangled state between many sensors requires extreme precision and control of experimental conditions, as well as advanced technologies for manipulating and measuring quantum systems. Despite these challenges, the enormous potential advantage of quantum sensors has stimulated intense theoretical and applied research. As we will see in the next section, these efforts have already led to significant progress, paving the way for new discoveries and applications that were once considered almost science fiction. Such developments not only promise to improve our ability to measure and observe the world but also offer a window into a deeper understanding of reality itself, according to the rules of the quantum world.

4.7 How to Build a Quantum Sensor

Let us take a closer look at the world of quantum sensors—a frontier that promises to significantly expand our ability to perceive the invisible. Unlike the arduous requirements needed to build a quantum computer, the engineering of a quantum sensor presents a much more accessible landscape. At the core of this technology is the ability to prepare a superposition of two states of any system—a requirement far less stringent than those for quantum computing. This opens the door to an incredibly wide range of possibilities.

As we have already mentioned, quantum sensors do not strictly require the use of entanglement, unlike quantum computing, where it is crucial. However, this does not mean that entanglement is not useful: as we have just seen, it is essential for pushing sensors toward reaching the Heisenberg limit and fully leveraging the quantum advantage, with the prospect of achieving precisions unimaginable just a few years ago.

Exploring the intricacies of building a quantum sensor, we encounter a variety of materials and architectures, each with its own unique characteristics and promises. We can classify sensors into two main families: on the one hand, there are sensors that rely on single atoms or a few fundamental elements, such as trapped ions or, as we will soon see, defects in diamonds; on the other hand, there are macroscopic sensors that, despite their size, retain surprising quantum properties.

4.7.1 *SQUIDs*

Among the most fascinating examples of this second category are SQUIDs (an acronym for Superconducting QUantum Interference Device) and Bose–Einstein condensates. A SQUID is an extraordinarily sensitive magnetic field detector, based on the same technology as the superconducting qubits we discussed in Chapter 2. In a SQUID, there is a superconducting ring through which current can flow freely. A time-varying magnetic field induces a circulating electric current in the ring. Thus, a SQUID can act as a sensor: if a magnetic field is turned on or varies, a current is generated inside it.

4.7.2 *Bose–Einstein condensates*

Bose–Einstein condensates are also macroscopic objects when compared to systems of a few atoms. They are clouds of tens of thousands of atoms cooled to temperatures near absolute zero—on the order of nanokelvin. A nanokelvin is one billionth of a Kelvin, where we have used the Kelvin scale, or absolute scale, to measure temperature. Zero Kelvin, or absolute zero, is the minimum possible temperature for any physical system and corresponds to −273.15°C. At absolute zero, the atoms composing the system would be "frozen" in the ground state—that is, they would possess the minimum possible energy.

Like superconductors, Bose–Einstein condensates are characterized by quantum properties of matter: the condensation from which they take their name means that all atoms "synchronize" and are described by the same properties. This state of matter was predicted by Albert Einstein and the Indian physicist Satyendra Nath Bose in the early 1920s and was first realized only at the end of the last century through experiments that earned the Nobel Prize in Physics in 2001 for Eric Cornell, Wolfgang Ketterle, and Carl Wieman. Normally, atoms have a velocity distribution centered around a value that depends on temperature, with great variability in the velocities of individual atoms. By lowering the temperature, this variability becomes increasingly small, to the point that—barring experimental uncertainties—the atoms have the same velocity (and furthermore "condense" into the same quantum state): this is a Bose–Einstein condensate.

Since the early pioneering experiments, the technology has spread worldwide, and it is now possible to obtain condensates on portable microchips. Condensates have also been created in space to study the effects of zero gravity. Furthermore, it is now possible to manipulate condensed atom clouds and create geometries that replicate those of SQUIDs, with which to build ultrasensitive sensors. Another possibility is to prepare the condensate in a superposition of two different states and measure its oscillations from one state to the other. These oscillations depend on an external magnetic field, and thus we have effectively created a sensor capable of measuring a magnetic field.

Both SQUIDs and Bose–Einstein condensates are considered quantum sensors because they leverage the quantum properties of the materials with which they are built (superconductivity in the first case and condensation in the second), but they are clearly special cases. In fact, being macroscopic, it is not easy to introduce entanglement between them (though there are efforts in this direction) and they do not achieve the performance of single-atom or few-atom quantum sensors.

4.7.3 *Color centers in diamonds*

Finally, we want to mention another possible implementation of quantum sensors that represents an advanced research area with revolutionary implications for numerous technological applications. The discovery that diamonds, beyond their remarkable beauty, also possess the ability to act as platforms for quantum sensors at room temperature has opened new prospects in various fields—from quantum computers to biomedicine. Let us explore some of the applications and potential of this emerging technology.

Natural diamonds are a crystalline structure of carbon atoms with fascinating properties. They appear transparent, but colored diamonds can also be found in nature: the color is due to the fact that within the regular structure of carbon atoms, during the diamond's formation, a small percentage of them are replaced by different atoms. In practice, there are defects in the regular crystalline structure, called color centers. Indeed, carbon is transparent (it does not interact with visible light), while other types of atoms may not be and can interact with visible light (absorbing and re-emitting it). Color centers thus act to make the light passing through the diamond colored.

All this is relevant for quantum sensors because the crystalline structure of diamond can be viewed as a perfect inert support for a single color center, which can be used as a sensor. What has actually been discovered is that, due to reasons related to the stability of the chemical structure, the color center is composed of a pair consisting of an atom other than carbon and a "vacancy"—that is, the absence of a carbon atom in one of the nearest positions of the crystalline lattice, as shown in Figure 4.1.

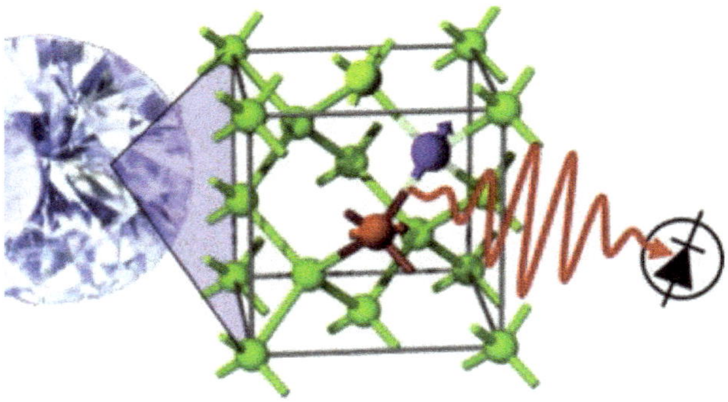

Fig. 4.1. Color center in diamond, formed by a defect (a nitrogen atom, red in the figure) and an adjacent vacancy (blue in the figure) in the crystalline structure of the diamond, which is composed almost exclusively of carbon atoms. The color center is used as a sensor. Figure taken from `https://www.sciencedirect.com/science/article/abs/pii/S0079656516300462`

The most studied of these is the NV (nitrogen-vacancy) color center, where a nitrogen atom and an adjacent vacancy create an artificial atom that can be exploited as a quantum sensor. Two of the energy levels of this artificial atom can indeed be used as a sensor following the scheme based on Rabi oscillations described earlier. The extraordinary thing is that it has been shown that these artificial atoms behave as quantum objects even at room temperature! They maintain quantum coherence, much like qubits in a quantum computer. In conclusion, colored defects in diamonds can be effectively used as quantum sensors, with application prospects enhanced by the fact that they can operate at room temperature.

Finally, it is worth noting that, fortunately, it is possible to create artificial diamonds and implant defects in them using various approaches—from specific growth techniques to methods in which impurities are first trapped and then fired one by one into the diamond with nanometer-scale precision (i.e., one impurity every few dozen atoms). It is therefore possible to envision preparing multiple defects in a global entangled state, with the aim of improving the precision of this type of sensor in the future. Some even propose this technology as a potential architecture for quantum computers. Furthermore, since diamond is a particularly inert material, it can be used in proximity to biological tissues, opening future opportunities in the biomedical field as well, as we will see in the next section.

4.8 From the Medicine of the Future to Gravitational Forces: The New Era of Quantum Sensors

Let us now provide an overview of some exciting potential applications of quantum sensors. First and foremost, quantum sensors are emerging as a revolutionary technology that promises to transform the landscape of medicine. These devices offer unprecedented precision and sensitivity in detecting biological and molecular signals. Applied to medicine, such sensors open the door to new diagnostic, therapeutic, and monitoring frontiers, anticipating a future where healthcare is personalized and extremely effective.

Let us start with sensors based on diamond defects. The biocompatibility of diamonds makes them ideal for biomedical applications: they can be used for real-time, high-precision monitoring of biological variables, in diagnostics, and even in the treatment of diseases through non-invasive techniques without damaging surrounding tissues. Quantum sensors can indeed be integrated into personalized monitoring devices that provide real-time data on a patient's health. Such devices enable the management of chronic conditions through constant monitoring of vital signs and specific biomarkers. The personalized approach offers doctors a detailed overview of a patient's condition, allowing for timely and tailored interventions.

One possible application is the use of diamond defect sensors as precision nanothermometers or pressure meters. In fact, changing the temperature of the diamond or the pressure on it causes its crystalline structure to deform slightly, and the end result is that the energy difference between the two energy levels defining the sensor will change slightly. We can therefore once again leverage Rabi oscillations to perform the measurement, whose frequency depends on the energy difference between the two levels and thus, ultimately, on the temperature (or pressure)—the quantity we want to measure.

The level of precision achievable and the fact that these sensors can be nanoscale with millidegree Celsius sensitivity have stimulated rapid technological development toward very ambitious goals—such as measuring the heat produced by chemical reactions between one or a few molecules, even inside living cells. In 2020, this technique was used in the laboratory to measure the internal temperature of some living worms. One might ask why it is interesting to measure the temperature of a worm, even going so far as to determine if it has a "fever". One reason is that the temperature inside a living organism is a measure of the biological activities taking place within the organism itself. Monitoring changes on a micrometer scale should provide detailed information about cellular and molecular activities. This could be important for clinical applications such as tissue structure imaging and the visualization of tumor heterogeneity. Since it is not easy to reduce the size of biocompatible thermometers to this scale, diamond defect sensors open fascinating prospects in this regard.

Sensors based on lattice defects in diamonds, such as the widely studied NV centers in recent years, can be used to detect changes in magnetic fields at the molecular level, enabling mapping of the internal structure of living cells without the use of harmful radiation. Finally, the sensitivity of quantum sensors allows for the detection of specific biomarkers even at low concentrations, enabling the identification of diseases such as cancer in their early stages. This could revolutionize screening practices, significantly improving cure rates through timely interventions.

Staying within the realm of possible biomedical applications, the United Kingdom has invested heavily in leveraging diamond defect sensors for brain diagnostics. Current sensors that enable brain diagnostics are highly bulky and thus do not allow monitoring during a patient's normal activities.

The Quantum Technology Hub—Sensing and Timing brings together various British universities and is one of the most active research centers in the development of these new quantum technologies. Recently, this center presented a new system based on quantum sensors to perform Magnetoencephalography (MEG)—a technique that enables monitoring of brain activity by measuring the magnetic fields produced by intracranial electric currents. This technique has been successfully used, for example, in identifying epileptic foci (i.e., the areas of the brain where seizures originate). However, an MEG machine is very complex, as it uses a large number of SQUID sensors (several hundred), which must be cooled to a temperature of 4 Kelvin (approximately −270°C) with liquid helium. In contrast, the system presented by the Quantum Technology Hub operates at room temperature and can be mounted on a special helmet that allows the patient to move almost normally. Looking ahead, it will be possible to perform analyses that monitor patients throughout a normal day: MEG is on its way to becoming a useful diagnostic tool with the same accessibility as the electrocardiogram, where patients are monitored for at least 24 consecutive hours using a Holter monitor (a portable device used to continuously check the electrical activity of the heart over an extended period) to identify certain cardiac conditions. Beyond diagnosis and monitoring, quantum sensors could play a crucial role in guided therapies. Their ability to detect specific molecular signals could be leveraged to deliver drugs in a targeted manner, reducing side effects and improving treatment effectiveness.

The field of quantum sensors based on diamond defects is still in the development phase, with research focused on improving sensitivity, miniaturization, and integration with other systems and devices. Challenges include increasing the quantum coherence of the sensors and integrating them with conventional electronics to create practical devices. The wide range of potential applications makes quantum sensors based on diamond defects one of the most promising areas of quantum physics and engineering.

Finally, we want to touch on the possibility of using quantum sensors to measure a fundamental force: gravity. Precisely measuring gravity has numerous technological, practical, and fundamental applications. For example, it is important for the correct synchronization of GPS (Global Positioning System) signals for satellite navigation positioning. GPS is based on a global network of artificial satellites orbiting approximately 20,000 kilometers from the Earth's surface that transmit radio signals to the Earth.

Einstein's theory of general relativity predicts that time passes at different speeds depending on the intensity of the gravitational field to which a clock is subjected. The closer a clock is to a large mass, such as the Earth, the stronger the gravitational field it experiences, and the slower time passes compared to a clock in a region with weaker gravity. This phenomenon is known as gravitational time dilation. Clocks on the Earth's surface are more influenced by the Earth's gravity than those on orbiting satellites. Consequently, clocks on Earth "actually run slower" than those on satellites. Without corrections, GPS satellite clocks would tend to drift by a few microseconds per day compared to terrestrial clocks—an error that translates to positioning inaccuracies of kilometers. To accurately calculate our position on Earth, our clock and those on the satellites must be precisely synchronized. Since light signals travel at 300,000 kilometers per second, a synchronization error of one millisecond would lead to an error of 300 kilometers in calculating our distance from a satellite. To achieve precisions on the order of a few meters, GPS systems take relativistic effects into account, ensuring that time measurements—and thus position determinations—are extremely accurate.

An emerging application of quantum sensors relates to navigation and geolocation. Quantum sensors have been developed to improve the precision of navigation systems. Unlike traditional GPS systems, which can be imprecise or unavailable in certain areas (such as underwater or in dense urban environments), quantum sensors based on interferometry can provide extremely precise localization even in these contexts. This technology promises to lead to significant developments for underwater navigation and autonomous vehicles.

Measuring gravitational force is also useful for the discovery and monitoring of oil reserves and aquifers. The gravitational force between two objects depends on their masses; therefore, if the mass of a first object is known, the mass of the second can be derived by measuring the gravitational force acting on it. By precisely measuring variations in gravitational force, we can determine if there is a cavity or liquid beneath us instead of solid rock. Clearly, to do this, the measurement must be precise enough to detect the difference between the Earth's gravitational force with or without the variable part (whether a tunnel, an oil reserve, or the underground magma of an active volcano). Recently, a quantum sensor successfully detected an underground tunnel one meter below the Earth's surface under normal external conditions (i.e., not in a controlled laboratory environment).

To build quantum gravitational sensors, the principle of superposition is always leveraged, constructing an interferometer. Simplifying greatly, atoms (for example, a Bose–Einstein condensate, which we discussed earlier) can be prepared to follow two different paths. Using electric fields, for example, it is possible to split an atomic beam like a river branching into two. After some time, through the reverse process, we recombine the two arms of the atomic beam. Unlike river water that splits, where each water molecule follows a single path, the experiment can be prepared such that in the recombined beam, each atom becomes a coherent quantum superposition of the two states corresponding to having passed through one path or the other. Once again, in the quantum world, instead of having a 50% probability of having passed through path A or path B, we can prepare a coherent superposition state of paths A and B—much like the Mona Lisa can be seen as a superposition of the happy and sad states of Mona Lisa. The fundamental point in this case is that the two possibilities correspond to two different trajectories: if the two paths are at different heights, the atoms will experience slightly different gravitational forces. This difference means that when the paths recombine, the state of the atoms depends on the difference in the gravitational field between the two trajectories—information that can be quantified with an appropriate measurement. We thus have an interferometer—that is, an instrument that measures the interference when the states of the atoms corresponding to two different paths are combined. The net result is that, using interference, it is possible to build gravitational sensors with exceptional performance.

Sensors of this type are already in use and even commercially available. For example, a quantum gravimeter (i.e., an instrument for measuring gravitational force) from the French company Muquans was placed on Mount Etna for a year to monitor its volcanic activity. Measuring variations in gravity around a volcano is important because it provides clues about variations in the density of underlying materials such as rock, gas, and magma. An increase in gravity, for example, could suggest the influx of a denser material like magma. The goal is to better understand what is happening inside the volcano, with the ambitious long-term objective of collecting sufficient data to predict eruptions.

The development and practical implementation of quantum gravitational sensors present significant challenges, including technical difficulties in maintaining atoms in superposition states for sufficiently long periods and the need for sophisticated and sensitive equipment to perform the measurements. However, continuous progress in quantum technologies and atom cooling techniques is leading to constant improvements in the precision and reliability of these extraordinary measuring instruments.

Chapter 5

The Quantum Horizon: The Challenges and Promises of a New World

> Science is an adventure of the whole human race to learn to live in and perhaps, to love the universe in which they are [...] It should be taught with a certain historical understanding, with a certain philosophical understanding, with a social understanding and a human understanding in the sense of the biography, the nature of the people who made this construction, the triumphs, the trials, the tribulations.
>
> Isidor Isaac Rabi, *Nobel Prize in Physics*

5.1 Living at the Forefront of Quantum Science: Challenges and Revolutions

Living in the community working in quantum sciences and technologies in recent years has been an exciting and stimulating experience. One October morning in 2022 marked a decisive turning point and most likely radically changed how this community of scientists has been perceived—and has perceived itself—since its birth about 40 years earlier. Those involved witnessed rapid developments and exciting prospects. However, despite the enthusiasm, quantum science research was often confined to a relatively isolated and marginal corner of the scientific landscape. The prevailing scientific interest in physics departments around the world focused on other research fields such as nuclear and subnuclear physics, astronomy, high-energy physics, and condensed matter physics.

Quantum sciences and technologies, for their part, had long since emerged from the obscure corners of research defined as "foundations of quantum mechanics", where a number of brilliant colleagues work to bring order to the basic aspects of quantum theory: What is the principle of reality? What is the profound meaning of entanglement? Why is time not quantized like space? These are some of the fascinating and engaging questions to which current science has not yet found definitive answers. Alternative theories to the quantum mechanics we know today have also been formulated. It is natural that many young researchers are drawn to these problems, even though they are often discouraged by older academics, who argue that devoting the best years of one's life to such research can be counterproductive for one's scientific career and a source of significant frustration.

Here emerges a broader, general dilemma that goes beyond the realm of quantum technologies or physics in general: Is it better for a young person to go with the flow of the most popular and accepted trends, or to embark on a path that, while more isolated, stimulates their curiosity and passion? Many are likely to ask themselves this question. One of the authors of this book remembers a conversation in the early 1970s—when he was a young researcher—at Princeton with Freeman Dyson, one of the greatest scientists of the last century, also known for his popular science works. Dyson advised avoiding the most crowded research fields, suggesting exploring new and emerging sectors where original discoveries are more likely, even with the risk of getting lost.

Returning to quantum sciences and technologies, on that October morning in 2022, the perspective changed. When the Royal Swedish Academy of Sciences' live-streamed press conference announced the Nobel Prize in Physics to Alain Aspect, John Clauser, and Anton Zeilinger "for experiments with entangled photons that have established the violation of Bell inequalities and opened the way to quantum information science," quantum sciences and technologies fully entered the pantheon of mainstream sciences. This recognition, combined with the enormous theoretical, experimental, and technological progress we have discussed in this book, has changed the perspective of the scientific community working on these topics, as well as that of several other communities. In fact, the development of quantum technologies will require not only physicists capable of deciphering fundamental quantum phenomena but also engineers, biologists, computer scientists, educators, business leaders, policymakers, and regulators who can implement and develop the necessary tools to grow this ecosystem in academia, industry, and ultimately in our lives.

5.2 Quantum Artificial Intelligence

5.2.1 *2022: The turning point*

Artificial intelligence, commonly known as AI, refers to the ability of computer systems to perform tasks that would normally require human intelligence. 2022 was a turning point not only for quantum technologies but also for AI, whose extraordinary potential became clear to the world. Tools like ChatGPT, which transform simple verbal instructions into images, texts, or videos, became accessible to the general public, triggering a true digital revolution. People who never would have imagined such possibilities were confronted with surprising creations that evoked feelings of wonder and, sometimes, sheer amazement.

However, the history of artificial intelligence dates back much further. We are in the 1940s of the last century, an era of scientific enthusiasm, when computing pioneers began conducting experiments with computers that sought to simulate self-organizing systems capable of performing simple tasks. It was then that the idea of "thinking" machines began to take shape.

We do not intend to recount the history of AI development here, but only to cite the emblematic case of the computer Deep Blue, which in 1996 managed to defeat the then world chess champion Garry Kasparov. This event caused a great stir at the time, and there was even talk of checkmate for human intelligence. At the time, AI development was viewed with mixed feelings, from optimism to dystopia. The latter vision was masterfully embodied by the computer HAL 9000, an artificial Polyphemus that in 2001: A Space Odyssey defeats humans at chess and later kills astronaut Frank Poole. In any case, Kasparov's defeat demonstrated AI's potential to tackle complex problems. Since its early days, AI has made giant strides, evolving into increasingly sophisticated systems capable of generating not only texts and images but also creative content, or managing our emails with surprising speed and efficiency. All this has been possible thanks to incredible progress in the computing power of supercomputers, which have enabled the optimization of neural networks. These are systems inspired by the functioning of the human brain, where millions of neurons communicate through synapses, exchanging electrical signals. In the digital world, a neural network learns and adapts in a similar way, through the optimization of millions of parameters that determine how the network responds to certain inputs, i.e., how the neural network answers the questions posed to it. Once these parameters are optimized (the learning phase), the neural network can perform various tasks: the more optimized parameters, the more powerful and versatile the systems. Today, with increasingly advanced networks, we can rely on AI systems that not only imitate some of our abilities but in some cases surpass them, creating high-quality texts or images in seconds.

The significance of artificial intelligence has been recognized by the Royal Swedish Academy of Sciences with the 2024 Nobel Prize in Physics awarded to John Hopfield and Geoffrey Hinton "for the fundamental discoveries and inventions that enable machine learning with artificial neural networks," the Nobel Prize in Chemistry to David Baker "for the computational design of proteins," and to Demis Hassabis and John Jumper for developing an AI model "for the prediction of protein structures."

Despite the fact that the basic algorithms and methods of neural networks have been known for decades, their practical application has been revolutionized by incredible advances in computing power. Thirty or forty years ago, computers were simply too slow to handle neural networks of significant size. But today, with the explosion of computing power and some crucial technical improvements, a world of possibilities has been unlocked that we could once only imagine. The importance of this computing power is so great that it now represents a critical and potentially limiting factor for the further development of artificial intelligence. Training a neural network requires it to learn from an enormous amount of data, and to do this, it must execute algorithms that require such large amounts of time and energy that they become, in fact, a limiting factor unless more efficient alternatives are found: Recent estimates[1] indicate that global energy consumption related to AI already equals that of a country like Ireland, and the costs to train a neural network and bring it to full potential can reach hundreds of millions of dollars.

In this scenario, where the demand for increasingly sophisticated artificial intelligence corresponds to a growing demand for computing power, it is natural to ask whether the new quantum computing paradigm we have learned about could be a possible solution. In 2014, Seth Lloyd of

[1] `https://spectrum.ieee.org/ai-energy-consumption`

the Massachusetts Institute of Technology and his collaborators provided a preliminary answer to this question, demonstrating that supervised machine learning—a basic form of AI that classifies data in a dataset—can be efficiently performed by a quantum computer.[2] This result not only signals a potential turning point in reducing AI's costs and energy consumption (since more efficient data classification would reduce computing time and thus ultimately energy expenditure) but could also represent the dawn of a new era in computing, where quantum computers will enable us to overcome the current limits of neural networks and open the way to even more extraordinary technological applications.

5.2.2 *How machines learn*

During the period when Seth Lloyd was laying the foundations for his innovative work at the Massachusetts Institute of Technology (MIT), the world of artificial intelligence seemed to barely touch that of quantum technologies. Most scientists tended to specialize exclusively in one of the two fields, rarely crossing paths with the other. Precisely during that time, one of the authors of this book was visiting MIT as a guest of Seth Lloyd and heard him almost every day discussing enthusiastically with his collaborators about new ideas related to artificial intelligence, without fully grasping their implications at the time—as often happens when one lacks the basic knowledge of a research field. In the following, we therefore aim to clarify what we mean today when we talk about "artificial intelligence", a term often shrouded in an aura of mystery and almost magic that should be dispelled so that everyone can understand the importance and real potential of AI.

Imagine living in a world without the ability to distinguish the objects around us: not knowing if what we see is red or green, an apple or a pear, a rose or a peony. This ability, which we take for granted, is fundamental for navigating and interacting effectively with our environment. But how can we teach a machine to "see" and recognize these details just as we do? This is where supervised learning comes into play, one of the earliest and most important techniques in the field of artificial intelligence. The goal is to create software, an algorithm, that can learn to classify and distinguish between two different categories of objects. For example, it could involve recognizing if an image shows a dog or a cat, or distinguishing between

[2]https://journals.aps.org/prl/abstract/10.1103/PhysRevLett.113.130503

two different numbers, like zero and one, or letters, like "a" and "z". This ability is not just an academic exercise; it is the foundation on which more complex skills like writing and reading are built. Supervised learning allows us to transfer part of our interpretive and decision-making abilities to machines. The possible applications are endless. From self-driving cars that operate without our intervention, reducing traffic accidents caused by human error, to diagnostic systems capable of early detection of diseases saving millions of lives, the applications of supervised learning are already beginning to shape a future where machines not only perform tasks but "collaborate" with us to create a safer and more efficient world. Supervised learning is therefore not only a fascinating field of study but also an essential key that can unlock revolutionary potential in technology, helping us create intelligent systems that improve our daily lives. But how exactly does it work? The process begins when AI developers "feed" their algorithms with enormous amounts of labeled data. Each piece of data acts like a teacher in a classroom: during the learning phase, the algorithm is shown the correct answer for that situation. Through this process, the machine "learns" to recognize patterns and make increasingly precise distinctions, step by step. Just as a child learns to recognize colors by repeatedly observing colored objects, a machine learns by identifying patterns in the data it receives. By showing it thousands of images of roads, it learns to independently distinguish between, for example, a green and a red traffic light. Similarly, by providing the algorithm with many examples of X-rays, it can learn to recognize signs of diseases, sometimes with a precision that surpasses that of human experts.

To delve deeper into how we teach machines to read, imagine going back to school, to the days when we learned to read and write. Remember when we tried to distinguish between handwritten letters, each with its small peculiarities? It is precisely from this basic experience that the world of machine learning draws inspiration—a technology that today allows us to teach machines to "read". Consider an apparently simple task: Distinguishing between two numbers, zero and one, written in cursive by many people—let's say 10,000 people, to be specific—with different handwriting. This challenge, although elementary, encapsulates the many variations that human handwriting can present. How can we teach a computer to do exactly this? The process begins by dividing a set of images into two groups. The first, composed of thousands of examples, is used to "train" the software. This phase is called "training" and works a bit like school lessons: We present our "digital student" with a series of images, asking it to identify if they show a zero or a one. If the software answers correctly we move on. If it makes a mistake, the algorithm corrects its internal parameters—just like a teacher corrects an assignment—ensuring that the machine "learns" from its error.

The second group of images is used for "testing", i.e., to evaluate how well the software has learned. If our software can correctly identify these new examples, we can confidently say that it "can read" the numbers zero to one in various handwriting styles—just like a child who, after learning at school, can read anything presented to them, except for the most illegible writing (like hastily written medical prescriptions!). The assumption we base this on is that, after training the machine on thousands of images, if it can give the correct answer for the test images, it will do the same for all subsequent ones we present to it. This learning method, which progressively refines response capabilities through error correction, is incredibly effective. It is the same principle that allows advanced technologies like ChatGPT to answer complex questions. It works because, given a sufficient number of parameters to adjust, the system can adapt to recognize extremely complex patterns—or, in technical terms, highly nonlinear functions. This is the beating heart of machine learning: A continuous learning process that, step by step, transforms machines into skilled interpreters of digital data, open to a universe of possibilities far beyond recognizing numbers on a screen.

5.2.3 *Quantum advantage?*

What Seth Lloyd and his collaborators, along with many other researchers, have examined is how to translate the process described above into a language that can adapt to a quantum computer, thus opening a window into a world of unexplored possibilities: The machine learning process, which traditionally relies on neural networks and complex algorithms, can be reformulated and optimized through the use of this type of computer. Quantum machine learning remains a field full of challenges, first and foremost due to the current limitations of quantum hardware. Today's real-world applications often rely on a small number of qubits and have not yet been able to surpass the capabilities of traditional computers. In fact, many of these applications have been explored by emulating quantum computers on classical machines. However, future prospects are promising: There are already clear ideas about how quantum machine learning could revolutionize the sector. Although it is likely that the most significant practical applications will only arrive in the medium or long term, this field continues to arouse enormous interest. For now, we must be patient, but the possibility of surpassing conventional artificial intelligence is real and intriguing.

It is interesting to note how the development of quantum algorithms has "inspired" innovative approaches to solving AI problems using classical computers. For years, neural networks have been the backbone of machine learning, but quantum computing has pushed toward the creation of tensor networks—mathematical structures designed to simulate quantum circuits on traditional computers. Surprisingly, in some cases, these tensor networks have proven more efficient than neural networks with the same resources, opening new possibilities for classical computing and offering a taste of the potential inspired by the quantum world.

As already mentioned, one of the critical problems in AI development is its enormous energy consumption, which has a significant impact on the environment. Quantum computers, however, have the potential to revolutionize this aspect: For the same computation, they tend to consume much less energy than traditional classical computers.[3] Although in the short term it is unlikely that quantum computers will radically transform the functionality of voice assistants or content generators (texts or images) that many of us use daily, their long-term impact could be very significant. In the future, artificial intelligence—increasingly integrated into our lives—could have a significantly reduced ecological footprint thanks to the efficiency of quantum computers.[4] This could represent a huge step forward toward a more sustainable world.

5.3 Reinventing Daily Life: How We Will Live in a Quantum World

In the following pages, we will see how quantum computers could influence our daily decisions. One of the most promising fields, especially for industrial applications, is operations research, where quantum computers could radically change the planning and optimization of resources, with direct impacts on our daily lives.

[3] `https://doi.org/10.1088/2058-9565/acae3e`

[4] `https://www.mckinsey.com/capabilities/mckinsey-digital/our-insights/quantum-computing-just-might-save-the-planet`

5.3.1 *Classical operations research*

Here's what it entails: Imagine facing common small problems—organizing household cleaning schedules considering everyone's commitments, packing luggage for vacation without forgetting anything, or choosing the most efficient route to run multiple errands in different places. These are all examples of operations research problems, a branch of mathematics that studies how to optimize complex organizational decisions. We often solve these tasks intuitively through years of accumulated experience, or with great patience, and they are often the subject of discussions between partners, siblings, or roommates.

In fact, it can be mathematically proven that these are all extremely difficult problems. So difficult that operations research was developed precisely to formalize them mathematically and then find the most efficient algorithms to solve them. Such problems have been studied and classified in their simplest but meaningful instances and are referred to as: personnel or resource management, the knapsack problem, the traveling salesman problem, etc. In the knapsack problem, given N objects—each characterized by a certain weight and a measure of how useful the object is—the goal is to choose which ones to put in the knapsack to achieve the minimum weight and maximum utility of the transported objects.

The traveling salesman problem, which we briefly discussed in Chapter 2, consists instead of finding the shortest route through various cities—a dilemma that anyone who has used a satellite navigator has probably faced in some way. Suppose the traveling salesman, starting from Padua, must visit Rome, Florence, Genoa, Venice, and Turin during his sales and promotion tour before returning to Padua. At this point, the algorithm must find the best solution, i.e., the route that visits all five cities with the total shortest distance traveled to save time and fuel. Furthermore, there may be other conditions to satisfy, for example, the traveling salesman must first pass through the port of Genoa to pick up goods that must then be delivered to Venice; or there may be tolls that in some cases he does not want to pay, or traffic jams (very likely at certain times) that he tries to avoid, etc. As the number of cities to visit increases, this problem becomes rapidly more complex, as the number of possible routes grows extremely quickly: For 5 cities, the number of different routes is 24; for 10 cities, it becomes 362,880; for 15 cities, the number exceeds 87 billion.

If we get in the car and ask the navigator to find a route with intermediate stops, we are asking the software to solve an optimization problem similar to the traveling salesman problem. And, as we all know, the software in our car or smartphone quickly proposes a solution. But are we sure that the solution provided is truly the best possible? If we think about it for a moment, answering this question involves not only knowing the solution given by the satellite navigator but also being able to compare it with all other possible ones.

Let's take the simplest case where, for each segment, there are only two possible roads to choose from, as schematically shown in Figure 5.1. For example, in our city, to get from work to pick up the children from school, we can take Via Aurelia or Via Barnaba (A or B for short). To also pass by the supermarket, we have two other possible routes, which for simplicity we will also denote as A or B. The possible combinations of the work–supermarket route via school are therefore AA, AB, BA, BB. If we complete the route home, also considering the two possible itineraries between the supermarket and our residence, the combinations become AAA, AAB, ABA, ABB, BAA, BAB, BBA, BBB. We already have eight different combinations: $2 \times 2 \times 2$.

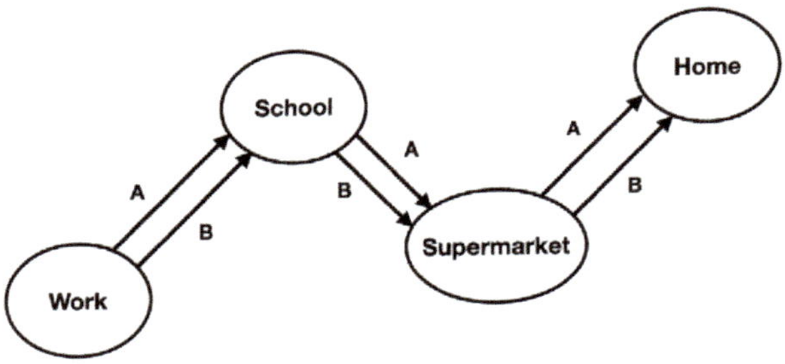

Fig. 5.1. Exponential growth of the number of possible work–home routes with the number of intermediate stops.

Suppose now that, when calculating a road route, a choice between two options must be made at each intersection or fork in the road. As we have seen, the number of possible combinations grows exponentially with the number of forks—so after 300 intersections, we reach 2 to the power of 300 (i.e., 2 multiplied by itself 300 times) different possibilities. The growth is the same as the legend of the pharaoh and the chessboard but with a chessboard having 300 squares instead of 64. The resulting number—2 to the power of 300—is larger than the number of atoms in the universe, making it practically impossible to examine each solution one by one to identify the best route.

The exponential growth of possible solutions and the impossibility of verifying that a proposed solution is indeed optimal are typical of decision-making problems. Problems like the "traveling salesman" are defined as combinatorial problems and are generally exponentially difficult.

We have emphasized that the explosion in the number of possible combinations to explore is the crux of the issue. But given the impossibility of verifying all possible solutions, how are these problems actually addressed? For decades, mathematicians and computer scientists have developed sophisticated algorithms to tackle them. Although it is not always possible to guarantee that the solution found is the absolute best, these algorithms can offer very good solutions—often sufficiently close to optimal. It is like in everyday life: We make choices based on experience, intuition, or even a bit of luck to navigate through complexity and chaos, such as in traffic on the way home after a long workday.

In the end, as we venture through the complexities of daily life with our strategies and choices—often without knowing if these choices are the best possible—operations research and its algorithms help us find order in chaos, suggesting paths that may not be perfect but allow us to move toward our goals.

Developers of algorithms for solving complex problems have drawn inspiration from methods developed by humanity over the centuries, as well as from nature's strategies. An example is the simplex method, which, starting from an initial position, seeks to follow the fastest path to the solution. Like when descending a mountain, at each fork we choose the path with the steepest descent. However, this strategy risks leading us into dead ends, from which we can escape by adding random steps during the search: In the analogy above, every now and then we take an upward path, thus exiting the dead end to resume the descent. Nature has also inspired revolutionary algorithms, such as genetic algorithms, which imitate the movements of flocks of birds hunting insects or introduce random elements—analogous to genetic mutations—into the search for problem solutions.

Finally, another strategy has been introduced from physics: annealing, or "tempering". It is a strategy inspired by the method used in metalworking and is used to find the optimal solution to combinatorial problems. This approach can be compared to a task many faced as children: organizing a box full of Lego pieces. There are basically two ways to sort Legos. The first method involves placing one piece at a time, neatly in the box, until it is completely full. Although systematic, this method has a major drawback: the time required grows proportionally to the number of pieces. For example, sorting 500 pieces will take 10 times as long as sorting 50.

However, there is an alternative that can be much more efficient: Throw all the pieces into the box and then shake it—initially vigorously and progressively more gently. The pieces begin to arrange themselves spontaneously, settling at the bottom and organizing into overlapping layers. This second strategy, although not perfect, often yields surprisingly good results—and more importantly, the time required to complete the task depends very little on the number of pieces. Similarly, there are algorithms that use annealing as a strategy to solve combinatorial problems, through seemingly unordered and random procedures—similar to the small shakes of the box—to find very effective solutions, even if not always the absolute best.

5.3.2 *Quantum operations research*

As we have inferred from these few introductory pages on operations research, the problems it seeks to solve are ubiquitous and of great importance for industrial processes and scientific research. The classical algorithms we use to solve them are numerous and constantly evolving.

In recent years, part of the scientific community and some industrial partners have begun to explore the use of quantum computers in this field. The motivation driving research in this direction is the fact that, as we have seen, the difficulty of combinatorial problems lies in the exponential growth of the number of possible configurations to explore. The attentive reader will not have missed the analogy with what was described in Chapter 1 about quantum registers: In that case too, the number of possible configurations of n quantum qubits grows exponentially with the number n of qubits. Thanks to the principle of quantum superposition, a quantum computer can exist in all these states simultaneously! Hence, the intuition that research can be accelerated by exploring all possible solutions simultaneously through quantum superposition.

Currently, practically every classical operations research algorithm has a quantum equivalent. We are witnessing a growing number of examples and formal demonstrations suggesting potential advantages in using quantum computers. Furthermore, the use of this technology is already showing practical benefits in various industrial contexts thanks to the technology and speed of superconducting qubits. In the following, we explore some of the most significant cases and potential applications.

The initial—and so far most impactful—push in this area has been made by the Canadian company D-Wave, which as early as 2011 proposed a superconducting qubit hardware system to solve a particular class of combinatorial problems through annealing, as described earlier. D-Wave is not a universal quantum computer capable of running the algorithms we discussed in Chapter 2, but it is still an excellent solver of combinatorial problems. To date, the company offers systems to solve combinatorial problems with thousands of variables that, in some cases, are competitive if not more performant than classical algorithms. Industrial applications include optimizing logistics at the Port of Los Angeles, optimizing Beijing taxi traffic from the airport to the city center, and much more.[5]

The applications of quantum computing to combinatorial optimization problems are demonstrating their potential in various sectors, ranging from logistics optimization to finance. For example, a group of IBM researchers, in collaboration with Delta Air Lines,[6] has successfully experimented with the use of quantum computers to optimize aircraft gate assignments, thereby reducing passenger transit times. Meanwhile, in Italy, Intesa Sanpaolo Bank[7] is exploring how quantum systems can be used to refine investment portfolio management strategies. A typical portfolio optimization problem—for which a quantum computer can be used—is choosing which assets to hold, sell, or buy to maximize expected return, given the level of risk we are willing to accept. In a different sector, the University of Padua, in collaboration with Thales Alenia Space Italy and on behalf of the European Space Agency, is using quantum methods to optimize the missions of artificial satellite constellations. These satellites are tasked with observing the Earth's surface, and the problem is to decide which trajectory and observations each satellite should make, considering constraints such as limited energy, time, memory, and communication bandwidth. Finally, in the energy and telecommunications sectors, large European and international companies are studying how quantum optimization can improve the management of energy networks—especially in the presence of renewable sources—or optimize the position of antennas for telecommunications repeaters.

---

[5]https://www.dwavesys.com/learn/featured-applications/

[6]https://arxiv.org/pdf/2111.09472.pdf

[7]https://group.intesasanpaolo.com/content/dam/portalgroup/repository-documenti/newsroom/mediabrief/230421_AbiLab_Quantum_Computing_in_Intesa_Sanpaolo.pdf

The examples cited highlight not only the versatility of quantum computing but also its growing impact on key industries and sectors. Many daily problems—both in industrial and social contexts—can be reduced to combinatorial problems and are currently addressed using classical algorithms. However, with rapid progress in the development of quantum hardware and software, we are approaching a turning point. As we will see in the following paragraph, if technological growth continues at its current pace, we may soon witness a significant impact of quantum technologies on many aspects of our lives.

In theory, a universal quantum computer is expected to solve operations research problems with unprecedented efficiency, optimizing routes and choices in fractions of the time required by classical computers. For example, in the traveling salesman problem, a quantum computer could simultaneously analyze all possible combinations of routes between cities like Rome, Florence, Padua, Turin, and Venice, and choose the most efficient one in terms of time and fuel. Even if there are additional variables—such as merchandise pickup times, tolls to avoid, or probable traffic jams—the quantum computer could integrate them into the analysis to find the optimal solution. A universal quantum computer could transform sectors such as logistics, resource management, and many others. Although quantum computers are still in their infancy and their practical applications require further development, the prospects are fascinating. We may not be far from the day when our daily choices are optimized thanks to the extraordinary power of quantum computing.

As the era of quantum computers approaches, numerous industries are beginning to prepare to harness the potential of this revolutionary technology. Combinatorial optimization problems—crucial for various sectors—are being reinterpreted in the new quantum language. Translating specific practical questions into this language, programming, and even conducting preliminary tests require highly specialized skills that are currently not easy to find. Large companies, aware of the opportunities that could open up with the arrival of reliable quantum computers with a large number of qubits, are motivated to stay at the forefront. Being prepared for the moment when advanced quantum computers may become available on the market is essential to maintain and increase competitive advantage. In this context, adapting quickly to the principles of quantum computing can make the difference between being able to lead innovation and being left behind.

5.4 Recent Developments and Future Horizons

In recent years, we have witnessed significant progress in the field of quantum computing, with the realization of new computer prototypes and the demonstration of increasingly advanced calculations that have often exceeded initial expectations. An example is the announcement of IBM's new chip (Condor, featuring 1121 qubits—though with limited performance in terms of precision)—which coincided with their latest industrial development roadmap published in December 2023. This chip represents an important qualitative leap: From initial prototypes with a small number of qubits (in 2016, IBM made a processor with only five qubits available to the public via remote access) and limited functionality, we have moved to much more complex systems. While efforts have so far focused on developing the hardware needed to support future chips, the next challenge will be to improve the performance of these superconducting chips. The reduced precision of the Condor chip is indeed a significant obstacle, as it limits the length of algorithms that can be effectively executed. In practice, the number of quantum gates that can be computed before errors compromise the reliability of results is still too low to perform truly useful calculations.

After developing a chip with 1000 qubits—and with the prospect of increasing this number in the coming years—IBM now faces a crucial challenge: Significantly improving the precision of individual logic gates and, above all, implementing effective error correction codes. In fact, the realization of a universal quantum computer—one that can solve a generic problem even when the complexity of the computation and the number of qubits required to solve it grow—depends decisively on the ability to correct errors that may arise during quantum calculations.

To achieve this goal, IBM plans to introduce a series of error mitigation algorithms, thereby increasing the precision of the results provided by their chips. By 2029, according to their roadmap, IBM expects to have developed chips that fully integrate error correction codes, enabling large-scale quantum calculations for practical and daily applications—such as those mentioned earlier in the operations research section. In the meantime, given the limited capabilities of current quantum computers, IBM has launched the "100×100 challenge",[8] an initiative inviting the discovery and development of algorithms with significant potential impact on science and industry. These algorithms must be compatible with a 100-qubit chip

[8] https://www.ibm.com/quantum/blog/next-wave-quantum-centric-supercomputing

(for now setting aside the Condor chip, which although it has over 1000 qubits, has limited precision), using approximately one hundred basic logic gates per qubit. The goal is to leverage currently available quantum chips to achieve unprecedented scientific results or industrial applications.

Regarding trapped ion-based quantum processors, they are known for their ability to perform logic operations on single qubits and two qubits with a very high level of precision—the best ever demonstrated in the laboratory—and have enabled extremely precise quantum calculations even on dozens of qubits. However, increasing the number of qubits with this technology remains a very complex issue. In fact, we are faced with a scalability problem: When trying to increase the number of qubits, the trap used to hold the ions in place and implement logic gates becomes progressively more unstable.

To solve this problem with trapped ion architecture, in 2023 the American company Quantinuum[9] took an important step by introducing the race track chip (see Figure 5.2). Quantinuum has implemented a solution to overcome scalability limits in trapped ion-based quantum computers. Instead of focusing on developing a single trap containing an increasing number of qubits, the company has designed a chip that houses many traps, which can communicate with each other by physically moving ions from one to another. In this way, each trap contains a limited number of ions, avoiding the stability problems we mentioned, while increasing the number of traps can in principle make the quantum computer scalable. Of course, this solution requires extremely sophisticated control of individual ions, including the ability to transport them and induce interactions between ions originating from different traps, in order to manipulate the overall quantum state of all ions. The race track chip realizes this vision—which just a few years ago would have seemed science fiction: Ions can be transferred along a circuit divided into two distinct zones. The "green zones" act as "memory areas", where ions reside when not directly involved in the execution of logic gates, maintaining the quantum state in which they were prepared. The "blue zones", on the other hand, are dedicated to executing the operations necessary for quantum computation: Initial state preparation, logic gates, and measurements. Another innovative aspect of this design is the ability to move ions in groups, for example, in groups of eight. This approach emulates the classical logic of bytes in the world of traditional computers, where a byte is composed of eight bits.

[9]`https://journals.aps.org/prx/pdf/10.1103/PhysRevX.13.041052`

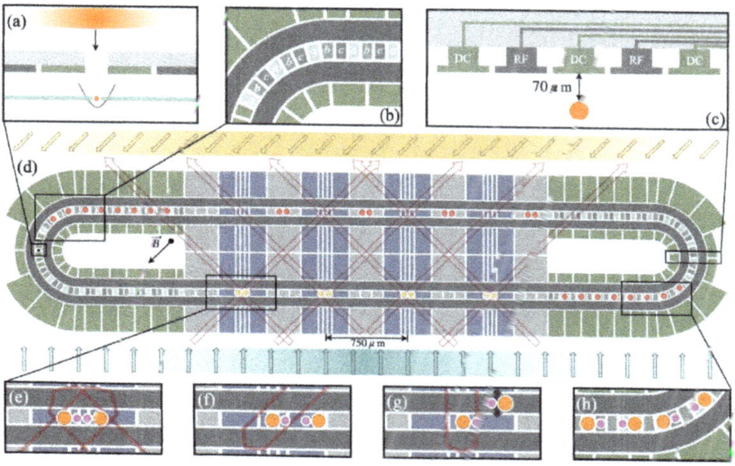

Fig. 5.2. Quantinuum's chip race track. Ions remain stationed in the green zones (the "memory areas" of the quantum computer), from which they are transferred into the blue zones to implement the logic gates required for quantum computation. Figure taken from `https://journals.aps.org/prx/abstract/10.1103/PhysRevX.13.041052`.

With the advent of this new technology featuring multiple communicating traps on a single chip, we have witnessed unexpected results—such as the demonstration of the creation of a 32-qubit entangled state (known as the GHZ state, in honor of physicists Daniel Greenberger, Michael Horne, and Anton Zeilinger). This success highlights the ability to generate highly complex quantum states with a large number of qubits. It not only demonstrates technical progress in controlling quantum systems but also opens the door to further developments that could make this architecture scalable, i.e., increasing the number of qubits available for computation. Within a few years, this could translate into the ability to demonstrate a "quantum advantage", solving certain problems (for example, simulating models relevant to understanding the magnetic and conductive properties of materials) with computing power superior to that of classical supercomputers.

Finally, while IBM presented its future plans (2023 roadmap), QuEra—a small and dynamic startup spun off from Harvard University—also presented its roadmap (see Figure 5.3) for the development of quantum computers based on Rydberg atoms (we remind you that these are atoms prepared in highly excited states to make them easier to manipulate). Thanks to collaboration with the research group of its co-founder, Harvard physics professor Mikhail Lukin, QuEra has promised significant progress—faster than what IBM proposed in its roadmap. In December 2023, QuEra demonstrated the ability to execute complex algorithms, correcting errors that occur during computation. To do this, it is necessary to introduce redundancy by increasing the number of qubits: For example, to perform a quantum computation with 48 qubits (called "logical" qubits, as they correspond to the number we actually have available for computation), QuEra used up to 280 qubits (called "physical" qubits, as they are the number that must be controlled and manipulated).[10] The reader should not be deceived by the fact that there are only 48 qubits here, while IBM's Condor chip has over 1000: The innovative aspect of QuEra's chip is its ability to correct errors—an essential element for then increasing the number of qubits and logic gates to obtain reliable results from quantum computation. The result achieved in Harvard's laboratory not only exceeded the expectations of the most optimistic supporters of the field by several years but also allowed the company to make credible promises (reported in Figure 5.3) that would otherwise have been considered unrealistic: A quantum computer available for commercial use with dozens, and soon hundreds, of qubits. If the company succeeds in realizing this roadmap, it will position itself as an undisputed leader in the sector, paving the way for new practical applications far beyond current market expectations.

Following QuEra's presentation of results, John Preskill of the California Institute of Technology declared that we are entering a new era: We have moved from the NISQ (Noisy Intermediate-Scale Quantum) era—characterized by still imperfect quantum computers with a limited number of qubits—to the LISQ (Logical Intermediate-Scale Quantum) era, where qubits are protected from errors and therefore much more reliable. Now, the challenge is to develop technology to increase the number of qubits, similar to what was done in previous decades with classical chips. The development of these chips has supported the incredible technological progress that has radically transformed our lives. It was an enormous mountain to climb, and now we must repeat the effort with this new technology. It will not be easy, but if we succeed, quantum algorithms will become accessible and can be used to solve complex problems in ways previously only imagined.

Around the world, many companies and startups are working on developing technologies similar to those discussed in these pages. Following are

[10] `https://www.nature.com/articles/s41586-023-06927-3`

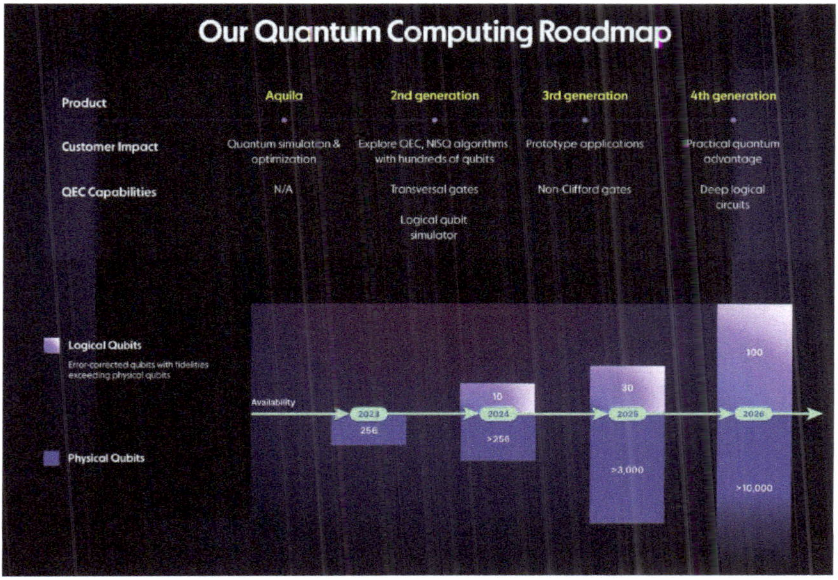

Fig. 5.3. Roadmap by the startup QuEra. The figure shows the projected development up to 2026, with the expectation of achieving by then a quantum advantage in problems of practical interest. Above the line, the "logical" qubits are shown as a function of the year; these are used to implement quantum algorithms and are protected from errors during computation through quantum error-correction (QEC) codes, which require redundancy and therefore use a much larger number of "physical" qubits, shown below the green line. Figure taken from `https://www.quera.com/qec`.

some examples related to Europe. The French startup PASQAL, based in Paris, is making significant progress with its Rydberg atom-based technology, positioning itself as a direct competitor to QuEra across the Atlantic. In Austria, Alpine Quantum Technologies—a spinoff of the University of Innsbruck and the Institute for Quantum Optics and Quantum Information (IQOQI) of the Austrian Academy of Sciences—continues to develop trapped ion technology. This company is based on a solid pioneering legacy of the University of Innsbruck, which realized two-qubit logic gates twenty years ago. Other companies in Germany are equally active, such as NEQXT and Planqc, which are exploring the potential of trapped ions and Rydberg atoms, respectively. These activities testify to the dynamism and growth potential of the quantum computing sector, promising important advancements for the future.

The development of quantum computing requires not only the development of hardware but also adequate software to program quantum computers, controlling and optimizing the various required operations (such as turning lasers or magnetic fields on and off, the duration and time modulation of a laser pulse, etc.). In the field of quantum computing software, tech giants like IBM, Google, and Amazon are leading development with a range of tools designed to program quantum algorithms. IBM offers Qiskit, Google has developed Cirq, and Amazon has launched Amazon Braket through its AWS (Amazon Web Services) arm, which enables access to cloud-based quantum computing services. Meanwhile, many startups and specialized companies are working closely with hardware manufacturers to refine existing solutions, improving both hardware and the control systems for electronic and optical components. For example, Quantum Machine and TOPTICA are known for their developments in control electronics and lasers. Finally, advanced software solutions are being developed to automate the control of quantum computers. An example is the German company Qruise, co-founded by one of the authors of this volume.

Finally, a growing number of startups are emerging with a focus on consulting and developing new algorithms for specific applications in quantum computing. Among these, the Spanish company Multiverse Computing specializes in financial applications, while the German company HQS is engaged in developing quantum simulations for new materials and chemical reactions.

As we can see, the quantum computing industry and related technologies go far beyond the specific issues related to quantum mechanics. Looking to the future, it is possible that the emerging field of quantum technologies will have an impact comparable to that of electronics and computing in previous decades. Consequently, such technologies could offer a range of opportunities for development and leadership in various industrial sectors to those who decide to invest and innovate today.

5.5 Conclusions

As we complete this book (in 2025), we still cannot know for sure if and when quantum computers and technologies will become part of everyday life. However, what we have recounted in these pages documents the incredible progress made: We have witnessed initially theoretical developments, followed by the rapid evolution of new technologies in academic and industrial laboratories. In many cases, it has been necessary not only to understand but also to invent and develop entirely new technologies—far beyond the originally intended applications.

Although the pace of development has been surprisingly fast, given the scale of the challenge, the road ahead is still long. Many obstacles remain—not only technological but also theoretical—such as the definition of more efficient error correction codes, the choice of optimal hardware for quantum computers (superconductors, atoms, ions, etc.), and the development of algorithms capable of solving increasingly complex problems with significant impact on sectors such as protein folding, pharmaceutical research, material modeling, and energy network optimization. These challenges promise to stimulate the formation of new research communities and professional figures who will combine scientific and industrial expertise to address them. As often happens, large-scale challenges open vast horizons of opportunities, ready to be seized in the coming years.

The long narrative of human history testifies that every significant discovery or invention has marked a decisive step forward for civilization. From the fundamental discovery of fire to the modern understanding of electromagnetic and nuclear forces, humans have constantly expanded their technological and scientific horizons. However, it is undeniable that alongside this remarkable progress, significant risks related to the misuse of such technologies have also emerged—with the danger of causing widespread harm. This risk is even greater today in an increasingly interconnected world dependent on technology itself.

The advancement of quantum technologies could, for example, lead to the development of extremely sensitive sensors capable of remotely monitoring the activity of individual neurons in our brains. This will enable an unprecedented understanding of the human brain's functioning, improving the treatment of neurological diseases. On the other hand, the ability to remotely monitor neuronal activity poses privacy risks, as brain activity is intimately linked to an individual's thoughts, feelings, and experiences. Furthermore, if sensors can not only monitor but also influence neurons, this could lead to dystopian scenarios where a person's thoughts and emotions could be controlled by external forces. This raises fundamental questions about individual freedom and autonomy. Similarly, quantum sensors—with their ability to detect extremely small variations in the environment—could be used to develop highly advanced surveillance systems. For example, sensors capable of detecting movements or activities through walls or other obstacles could be exploited to monitor people without their consent. These risks are compounded by the fact that quantum technologies require significant investments in research and development, which could lead to an even wider technological gap between countries or private corporations that can afford these technologies and others that are excluded. This is a common theme with artificial intelligence that should be a priority for a country's leaders.

Currently, there is an active debate about the limits of a machine and whether computers will become increasingly intelligent—eventually being able to emulate the human brain. Quantum computers could do this with much greater efficiency than classical computers. This raises ethical questions about the nature of consciousness and identity. If a machine could replicate the functioning of the human brain, we would have to ask whether it possesses some form of consciousness. Meanwhile, robots—or rather sensors—will no longer only inhabit factories or cars but will also begin to inhabit us, changing us. In essence, the boundaries between what is human and what is machine could become more blurred, triggering questions about the ethical limits of altering or enhancing the human body with technology.

Faced with this reality, the importance of solid and widespread scientific education becomes increasingly evident. Every person should be able to understand scientific information just as they can read and write. This is not only a personal enrichment but a collective necessity: An educated and well-informed population is fundamental both to benefit from the advantages and to manage the risks associated with emerging technologies, and to navigate the ethical and responsible use of scientific discoveries. Our hope is that this volume can also contribute to this direction.

Index

www.ingramcontent.com/pod-product-compliance
Ingram Content Group UK Ltd.
Pitfield, Milton Keynes, MK11 3LW, UK
UKHW062101120926
8588IPUK00019B/196

9 789819 830879